CAMBRIDGE TRACTS IN MATHEMATICS

General Editors

---

165 Projective Differential Geometry Old and New

Projective Differential Geometry Old and New

From the Schwarzian Derivative to Cohomology of Diffeomorphism Groups

Ideas of projective geometry keep reappearing in seemingly unrelated fields of mathematics. In this book, the authors' main goal is to emphasize connections between classical projective differential geometry and contemporary mathematics. They also give new results and new proofs of classic theorems. Exercises play a prominent role: historical and cultural comments relate the basic notions to a broader context.

The book starts with a detailed discussion of the simplest differential projective invariant, the Schwarzian derivative, and its connection to the Virasoro algebra, one of the most popular objects of study in mathematical physics today. About half of the book is devoted to one-dimensional projective differential geometry and related topics: differential operators, the cohomology of the group of diffeomorphisms of the circle, the classical four-vertex theorem, its numerous generalizations and their significance to symplectic and contact topology. The classical projective differential geometry of projective hypersurfaces is surveyed and related to less traditional topics such as the complete integrability of the geodesic flow on the ellipsoid, Hilbert's fourth problem, and very recent results and conjectures in the projective differential topology of surfaces. A final chapter considers various versions of the multi-dimensional Schwarzian derivative, including the Lagrangian Schwarzian and multi-dimensional Sturm theory. The seven appendices cover diverse background material from proofs of the Sturm–Hurwitz theorem to symplectic and contact geometry, from the Godbillon–Vey class to infinite-dimensional Poisson geometry.

In sum, this book is a rapid route for graduate students and researchers to the frontiers of current research in this evergreen subject.

V. Ovsienko
Université Claude Bernard Lyon 1

S. Tabachnikov
Pennsylvania State University

Projective Differential Geometry Old and New

From the Schwarzian Derivative to the Cohomology of Diffeomorphism Groups

CAMBRIDGE
UNIVERSITY PRESS

CAMBRIDGE UNIVERSITY PRESS
Cambridge, New York, Melbourne, Madrid, Cape Town, Singapore,
São Paulo, Delhi, Dubai, Tokyo, Mexico City

Cambridge University Press
The Edinburgh Building, Cambridge CB2 8RU, UK

Published in the United States of America by Cambridge University Press, New York

www.cambridge.org
Information on this title: www.cambridge.org/9780521831864

First published 2005

A catalogue record for this publication is available from the British Library

Library of Congress Cataloging in Publication data

Ovsienko, Valentin.
Projective differential geometry old and new: from the Schwarzian derivative to the cohomology of diffeomorphism groups / V. Ovsienko, S. Tabachnikov.
p. cm. – (Cambridge tracts in mathematics; 165)
Includes bibliographical references and index.
ISBN 0 521 83186 5 (hardback)
1. Projective differential geometry. I. Tabachnikov, Serge. II. Title. III. Series.

QA660.O87 2005
516.36–dc22 2004045919

ISBN 978-0-521-83186-4 Hardback

Contents

Preface: why projective?

> Metrical geometry is a part of descriptive geometry,[1] and descriptive geometry is all geometry.
>
> *Arthur Cayley*

On October 5, 2001, the authors of this book typed in the word "Schwarzian" in the MathSciNet database and the system returned 666 hits. Every working mathematician has encountered the Schwarzian derivative at some point of his education and, most likely, tried to forget this rather scary expression right away. One of the goals of this book is to convince the reader that the Schwarzian derivative is neither complicated nor exotic; in fact, it is a beautiful and natural geometrical object.

The Schwarzian derivative was discovered by Lagrange: "According to a communication for which I am indebted to Herr Schwarz, this expression occurs in Lagrange's researches on conformable representation 'Sur la construction des cartes géographiques'" [117]; the Schwarzian also appeared in a paper by Kummer in 1836, and it was named after Schwarz by Cayley. The main two sources of current publications involving this notion are classical complex analysis and one-dimensional dynamics. In modern mathematical physics, the Schwarzian derivative is mostly associated with conformal field theory. It also remains a source of inspiration for geometers.

The Schwarzian derivative is the simplest projective differential invariant, namely, an invariant of a real projective line diffeomorphism under the natural $\mathrm{SL}(2,\mathbb{R})$-action on $\mathbb{RP}^1$. The unavoidable complexity of the formula for the Schwarzian is because $\mathrm{SL}(2,\mathbb{R})$ is such a large group (a three-dimensional symmetry group of a one-dimensional space).

[1] By descriptive geometry Cayley means projective geometry; this term was in use in the mid nineteenth century.

Projective geometry is simpler than affine or Euclidean geometries: in projective geometry, there are no parallel lines or right angles, and all non-degenerate conics are equivalent. This shortage of projective invariants is due to the fact that the group of symmetries of the projective space $\mathbb{RP}^n$ is large. This group, $\mathrm{PGL}(n+1, \mathbb{R})$, is equal to the quotient of $\mathrm{GL}(n+1, \mathbb{R})$ by its center. The greater the symmetry group, the fewer invariants it has. For instance, there exists no $\mathrm{PGL}(n+1, \mathbb{R})$-invariant tensor field on $\mathbb{RP}^n$, such as a metric or a differential form. Nevertheless, many projective invariants have been found, from the Ancient Greeks' discovery of configuration theorems to differential invariants. The group $\mathrm{PGL}(n+1, \mathbb{R})$ is maximal among Lie groups that can act effectively on n-dimensional manifolds. It is because of this maximality that projective differential invariants, such as the Schwarzian derivative, are uniquely determined by their invariance properties.

Projective geometry used to be a core subject in the university curriculum and, as late as the first half of the twentieth century, projective differential geometry was at the cutting edge of geometric research. Nowadays this subject occupies a more modest position, and it is rare for a mathematics major to be familiar with the Pappus or Desargues theorems.

This book is not an exhaustive introduction to projective differential geometry or a survey of its recent developments. It is addressed to the reader who wishes to cover a greater distance in a short time and arrive at the front line of contemporary research. This book can serve as a basis for graduate topics courses. Exercises play a prominent role while historical and cultural comments relate the subject to a broader mathematical context. Parts of this book have been used for topic courses and expository lectures for undergraduate and graduate students in France, Russia and the USA.

Ideas of projective geometry keep reappearing in seemingly unrelated fields of mathematics. The authors of this book believe that projective differential geometry is still very much alive and has a wealth of ideas to offer. Our main goal is to describe connections between classical projective geometry and contemporary research and thus to emphasize the unity of mathematics.

Acknowledgments. For many years we have been inspired by our teachers V. I. Arnold, D. B. Fuchs and A. A. Kirillov, who made a significant contribution to the modern understanding of the material of this book. It is a pleasure to thank our friends and collaborators C. Duval, B. Khesin, P. Lecomte and C. Roger, whose many results are included here. We are much indebted to J. C. Alvarez, M. Ghomi, E. Ghys, J. Landsberg, S. Parmentier, B. Solomon, G. Thorbergsson and M. Umehara for enlightening discussions and help. It was equally pleasant and instructive to work with our younger colleagues and students S. Bouarroudj,

H. Gargoubi, L. Guieu and S. Morier-Genoud. We are grateful to the Shapiro Fund at Penn State, the Research in Pairs program at Oberwolfach and the National Science Foundation for their support.

1

Introduction

...the field of projective differential geometry is so rich that it seems well worth while to cultivate it with greater energy than has been done heretofore.

E. J. Wilczynski

In this introductory chapter we present a panorama of the subject of this book. The reader who decides to restrict himself to this chapter will get a rather comprehensive impression of the area.

We start with the classical notions of curves in projective space and define projective duality. We then introduce first differential invariants such as projective curvature and projective length of non-degenerate plane projective curves. Linear differential operators in one variable naturally appear here to play a crucial role in what follows.

Already in the one-dimensional case, projective differential geometry offers a wealth of interesting structures and leads us directly to the celebrated Virasoro algebra. The Schwarzian derivative is the main character here. We tried to present classical and contemporary results in a unified synthetic manner and reached the material discovered as late as the last decades of the twentieth century.

1.1 Projective space and projective duality

Given a vector space V, the associated projective space, $\mathbb{P}(V)$, consists of one-dimensional subspaces of V. If $V = \mathbb{R}^{n+1}$ then $\mathbb{P}(V)$ is denoted by $\mathbb{RP}^n$. The projectivization, $\mathbb{P}(U)$, of a subspace $U \subset V$ is called a projective subspace of $\mathbb{P}(V)$.

The dual projective space $\mathbb{P}(V)^*$ is the projectivization of the dual vector space V^*. Projective duality is a correspondence between projective subspaces of $\mathbb{P}(V)$ and $\mathbb{P}(V)^*$, the respective linear subspaces of V and V^* are annulators of each other. Note that projective duality reverses the incidence relation.

Natural local coordinates on $\mathbb{RP}^n$ come from the vector space $\mathbb{R}^{n+1}$. If $x_0, x_1, \ldots, x_n$ are linear coordinates in $\mathbb{R}^{n+1}$, then $y_i = x_i/x_0$ are called *affine coordinates* on $\mathbb{RP}^n$; these coordinates are defined in the chart $x_0 \neq 0$. Likewise, one defines affine charts $x_i \neq 0$. The transition functions between two affine coordinate systems are linear-fractional.

Projectively dual curves in dimension 2

The projective duality extends to curves. A smooth curve γ in $\mathbb{RP}^2$ determines a one-parameter family of its tangent lines. Each of these lines gives a point in the dual plane $\mathbb{RP}^{2^*}$ and we obtain a new curve γ^* in $\mathbb{RP}^{2^*}$, called the *dual curve*.

In a generic point of γ, the dual curve is smooth. Points in which γ^* has singularities correspond to *inflection* of γ. In generic points, γ has order 1 contact with its tangent line; inflection points are those points where the order of contact is higher.

Exercise 1.1.1 Show that:
(a) two parabolas, given in affine coordinates by $y = x^\alpha$ and $y = x^\beta$, are dual for $1/\alpha + 1/\beta = 1$;
(b) the curves in Figure 1.2 are dual to each other.

A fundamental fact is that $(\gamma^*)^* = \gamma$, which justifies the terminology (a proof is given in the next subsection). As a consequence, one has an alternative

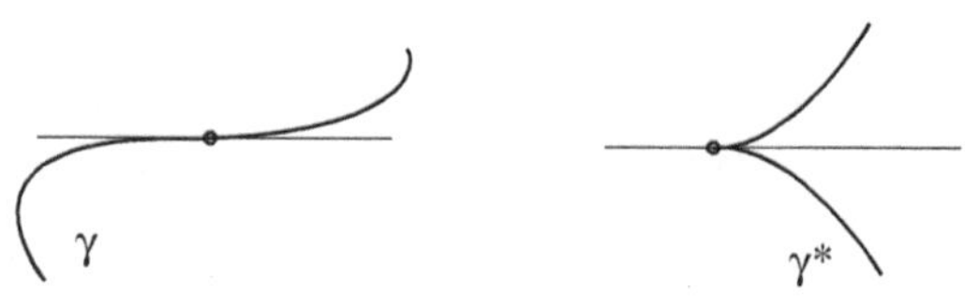

Fig. 1.1. Duality between an inflection and a cusp

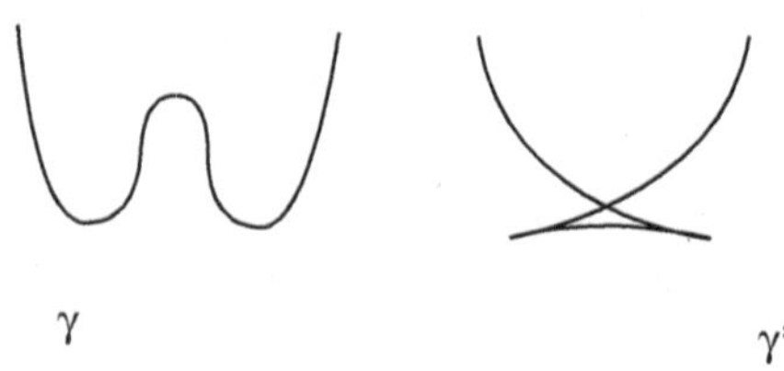

Fig. 1.2. Projectively dual curves

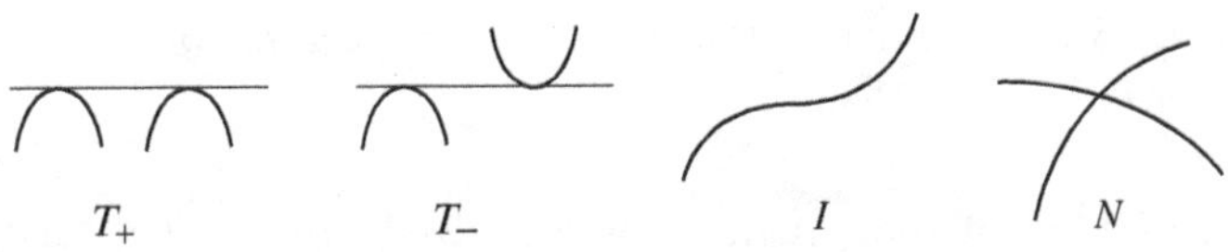

Fig. 1.3. Invariants of plane curves

definition of the dual curve. Every point of γ determines a line in the dual plane, and the envelope of these lines is γ^*.

Two remarks are in order. The definition of the dual curve extends to curves with cusps, provided the tangent line is defined at every point and depends on the point continuously. Secondly, duality interchanges double points with double tangent lines.

Exercise 1.1.2 Consider a generic smooth closed immersed plane curve γ. Let $T_\pm$ be the number of double tangent lines to γ such that locally γ lies on one side (respectively, opposite sides) of the double tangent (see Figure 1.3), I be the number of inflection points and N the number of double points of γ. Prove that

$$T_+ - T_- - \tfrac{1}{2} I = N.$$

Hint. Orient γ and let $\ell(x)$ be the positive tangent ray at $x \in \gamma$. Consider the number of intersection points of $\ell(x)$ with γ and investigate how this number changes as x traverses γ. Do the same with the negative tangent ray.

Projective curves in higher dimensions

Consider a generic smooth parameterized curve $\gamma(t)$ in $\mathbb{RP}^n$ and its generic point $\gamma(0)$. Construct a flag of subspaces as follows. Fix an affine coordinate system, and define the *k-th osculating subspace* F_k as the span of $\gamma'(0), \gamma''(0), \dots, \gamma^{(k)}(0)$. This projective space depends neither on the parameterization nor on the choice of affine coordinates. For instance, the first osculating space is the tangent line; the $(n-1)$-th is called the osculating hyperplane.

A curve γ is called *non-degenerate* if, in every point of γ, one has the full osculating flag

$$F_1 \subset \cdots \subset F_n = \mathbb{RP}^n. \tag{1.1}$$

A non-degenerate curve γ determines a 1-parameter family of its osculating hyperplanes. Each of these hyperplanes gives a point in the dual space $\mathbb{RP}^{n*}$, and we obtain a new curve γ^* called the dual curve.

As before, one has the next result.

Theorem 1.1.3 *The curve that is dual to a non-degenerate one is smooth and non-degenerate, and* $(\gamma^*)^* = \gamma$.

Proof Let $\gamma(t)$ be a non-degenerate parameterized curve in $\mathbb{RP}^n$, and $\Gamma(t)$ its arbitrary lift to $\mathbb{R}^{n+1}$. The curve $\gamma^*(t)$ lifts to a curve $\Gamma^*(t)$ in the dual vector space satisfying the equations

$$\Gamma \cdot \Gamma^* = 0, \quad \Gamma' \cdot \Gamma^* = 0, \quad \ldots, \quad \Gamma^{(n-1)} \cdot \Gamma^* = 0, \tag{1.2}$$

where dot denotes the pairing between vectors and covectors. Any solution $\Gamma^*(t)$ of (1.2) projects to $\gamma^*(t)$. Since γ is non-degenerate, the rank of system (1.2) equals n. Therefore, $\gamma^*(t)$ is uniquely defined and depends smoothly on t.

Differentiating system (1.2), we see that $\Gamma^{(i)} \cdot \Gamma^{*(j)} = 0$ for $i + j \leq n - 1$. Hence the osculating flag of the curve γ^* is dual to that of γ and the curve γ^* is non-degenerate. In particular, for $i = 0$, we obtain $\Gamma \cdot \Gamma^{*(j)} = 0$ with $j = 0, \ldots, n-1$. Thus, $(\gamma^*)^* = \gamma$. □

As in the two-dimensional case, the dual curve γ^* can be also obtained as the envelope of a 1-parameter family of subspaces in $\mathbb{RP}^{n*}$, namely, of the dual k-th osculating spaces of γ. All this is illustrated by the following celebrated example.

Example 1.1.4 Consider a curve $\gamma(t)$ in $\mathbb{RP}^3$ given, in affine coordinates, by the equations:

$$y_1 = t, \quad y_2 = t^2, \quad y_3 = t^4.$$

This curve is non-degenerate at point $\gamma(0)$. The plane dual to point $\gamma(t)$ is given, in an appropriate affine coordinate system (a_1, a_2, a_3) in $\mathbb{RP}^{3*}$, by the equation

$$t^4 + a_1 t^2 + a_2 t + a_3 = 0. \tag{1.3}$$

This one-parameter family of planes envelops a surface called *the swallow tail* and shown in Figure 1.4. This developable surface consists of the tangent lines to the curve γ^*. Note the cusp of γ^* at the origin.

Comment

The study of polynomials (1.3) and Figure 1.4 go back to the nineteenth century [118]; the name "swallow tail" was invented by R. Thom in the mid twentieth century in the framework of the emerging singularity theory (see [16]). The swallow tail is the set of polynomials (1.3) with multiple roots, and the curve γ^* corresponds to polynomials with triple roots. This surface is a typical example of

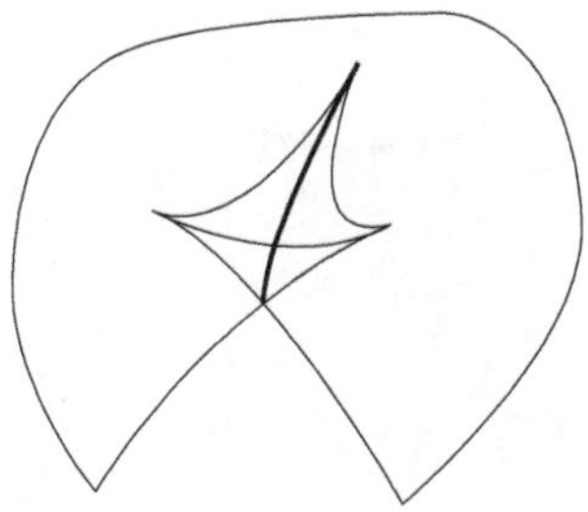

Fig. 1.4. Swallow tail

a developable surface, i.e., a surface of zero Gauss curvature. The classification of developable surfaces is due to L. Euler (cf. [193]): generically, such a surface consists of the tangent lines of a curve, called the edge of regression. The edge of regression itself has a singularity as in Figure 1.4.

Unlike the Plücker formula of classic algebraic geometry, the result of Exercise 1.1.2 is surprisingly recent; it was obtained by Fabricius-Bjerre in 1962 [61]. This result has numerous generalizations; see, for example, [199, 66].

1.2 Discrete invariants and configurations

The oldest invariants in projective geometry are projective invariants of configurations of point and lines. Our exposition is just a brief excursion into the subject; for a thorough treatment see, e.g., [22].

The cross-ratio

Consider the projective line $\mathbb{RP}^1$. Every triple of points can be taken to any other triple by a projective transformation. This is not the case for quadruples of points: four points in $\mathbb{RP}^1$ have a numeric invariant called the *cross-ratio*. Choosing an affine parameter t to identify $\mathbb{RP}^1$ with $\mathbb{R} \cup \{\infty\}$, the action of $\mathrm{PGL}(2, \mathbb{R})$ is given by linear-fractional transformations:

$$t \mapsto \frac{at+b}{ct+d}. \tag{1.4}$$

The four points are represented by numbers t_1, t_2, t_3, t_4, and the cross-ratio is defined as

$$[t_1, t_2, t_3, t_4] = \frac{(t_1 - t_3)(t_2 - t_4)}{(t_1 - t_2)(t_3 - t_4)}. \tag{1.5}$$

A quadruple of points is called *harmonic* if its cross-ratio is equal to -1.

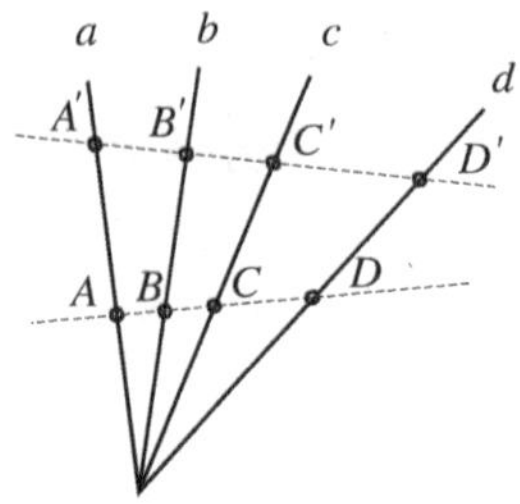

Fig. 1.5. Cross-ratio of lines: $[A, B, C, D] = [A', B', C', D'] := [a, b, c, d]$

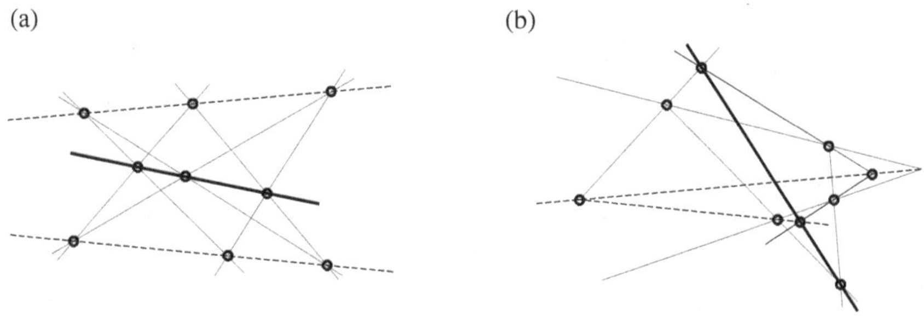

Fig. 1.6. Pappus and Desargues theorems

Exercise 1.2.1 (a) Check that the cross-ratio does not change under transformations (1.4).
(b) Investigate how the cross-ratio changes under permutations of the four points.

One defines also the cross-ratio of four concurrent lines in $\mathbb{RP}^2$, that is, four lines through one point. The pencil of lines through a point identifies with $\mathbb{RP}^1$, four lines define a quadruple of points in $\mathbb{RP}^1$, and we take their cross-ratio. Equivalently, intersect the four lines with an auxiliary line and take the cross-ratio of the intersection points therein, see Figure 1.5.

Pappus and Desargues

Note two configurations in the projective plane. Figure 1.6 depicts two classical theorems.

The Pappus theorem describes the following construction which we recommend to the reader to perform using a ruler or his favorite drawing software. Start with two lines, pick three points on each. Connect the points pairwise as shown in Figure 1.6a to obtain three new intersection points. These three points are also collinear.

In the Desargues theorem, draw three lines through one point and pick two points on each to obtain two perspective triangles. Intersect the pairs of corresponding sides of the triangles. The three points of intersection are again collinear, as shown in Figure 1.6b.

Pascal and Brianchon

The next theorems, depicted in Figure 1.7, involve conics. To obtain the Pascal theorem, replace the two original lines in the Pappus configuration by a conic. In the Brianchon theorem, circumscribe a hexagon about a conic and connect the opposite vertices by diagonals. The three lines intersect at one point.

Unlike the Pappus and Desargues configurations, the Pascal and Brianchon ones are projectively dual to each other.

Steiner

Steiner's theorem provides a definition of the cross-ratio of four points on a conic. Choose a point P on a conic. Given four points A, B, C, D, define their cross-ratio as that of the lines (PA), (PB), (PC), (PD). The theorem asserts that this cross-ratio is independent of the choice of point P:

$$[(PA), (PB), (PC), (PD)] = [(P_1A), (P_1B), (P_1C), (P_1D)]$$

in Figure 1.8.

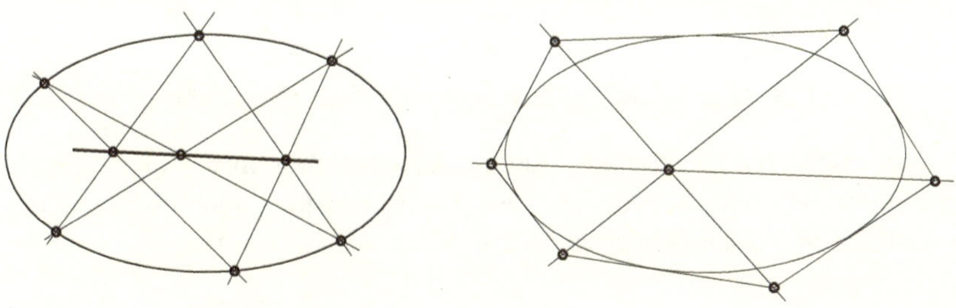

Fig. 1.7. Pascal and Brianchon theorems

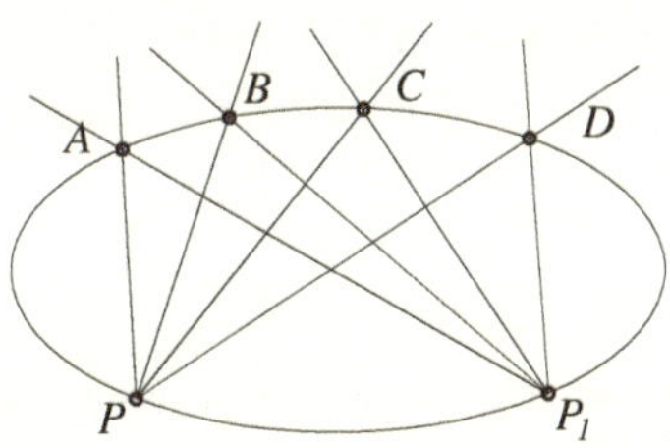

Fig. 1.8. Steiner theorem

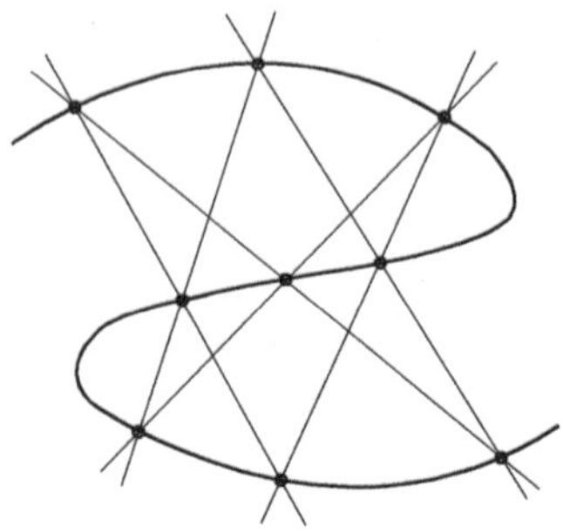

Fig. 1.9. Generalized Pappus theorem

Comment

In 1636 Girard Desargues published a pamphlet "A sample of one of the general methods of using perspective" that laid the foundation of projective geometry; the Desargues theorem appeared therein. The Pappus configuration is considerably older; it was known as early as the third century AD. The triple of lines in Figure 1.6a is a particular case of a cubic curve; the Pappus configuration holds true for nine points on an arbitrary cubic curve – see Figure 1.9. This more general formulation contains the Pascal theorem as well. Particular cases of Steiner's theorem were already known to Apollonius.[1] Surprisingly, even today, there appear new generalizations of the Pappus and the Desargues theorems; see [182, 183].

1.3 Introducing the Schwarzian derivative

Projective differential geometry studies projective invariants of functions, diffeomorphisms, submanifolds, etc. One way to construct such invariants is to investigate how discrete invariants vary in continuous families.

The Schwarzian derivative and cross-ratio

The best known and most popular projective differential invariant is the Schwarzian derivative. Consider a diffeomorphism $f : \mathbb{RP}^1 \to \mathbb{RP}^1$. The Schwarzian derivative measures how f changes the cross-ratio of infinitesimally close points.

[1] We are indebted to B. A. Rosenfeld for enlightening discussions on Ancient Greek mathematics.

Let x be a point in $\mathbb{RP}^1$ and v be a tangent vector to $\mathbb{RP}^1$ at x. Extend v to a vector field in a vicinity of x and denote by ϕ_t the corresponding local one-parameter group of diffeomorphisms. Consider four points:

$$x, \quad x_1 = \phi_\varepsilon(x), \quad x_2 = \phi_{2\varepsilon}(x), \quad x_3 = \phi_{3\varepsilon}(x)$$

(ε is small) and compare their cross-ratio with that of their images under f. It turns out that the cross-ratio does not change in the first order in ε:

$$[f(x), f(x_1), f(x_2), f(x_3)] = [x, x_1, x_2, x_3] - 2\varepsilon^2 S(f)(x) + O(\varepsilon^3). \tag{1.6}$$

The ε^2-coefficient depends on the diffeomorphism f, the point x and the tangent vector v, but not on its extension to a vector field.

The term $S(f)$ is called the Schwarzian derivative of a diffeomorphism f. It is homogeneous of degree 2 in v and therefore $S(f)$ is a *quadratic differential* on $\mathbb{RP}^1$, that is, a quadratic form on $T\mathbb{RP}^1$.

Choose an affine coordinate $x \in \mathbb{R} \cup \{\infty\} = \mathbb{RP}^1$. Then the projective transformations are identified with linear-fractional functions, and quadratic differentials are written as $\phi = a(x)\,(\mathrm{d}x)^2$. The change of variables is then described by the formula

$$\phi \circ f = \left(f'\right)^2 a(f(x))\,(\mathrm{d}x)^2. \tag{1.7}$$

The Schwarzian derivative is given by the formula

$$S(f) = \left(\frac{f'''}{f'} - \frac{3}{2}\left(\frac{f''}{f'}\right)^2\right)(\mathrm{d}x)^2. \tag{1.8}$$

Exercise 1.3.1 (a) Check that (1.6) contains no term linear in ε.
(b) Prove that $S(f)$ does not depend on the extension of v to a vector field.
(c) Verify formula (1.8).

The Schwarzian derivative enjoys remarkable properties.

- By the very construction, $S(g) = 0$ if g is a projective transformation, and $S(g \circ f) = S(f)$ if g is a projective transformation. Conversely, if $S(g) = 0$ then g is a projective transformation.
- For arbitrary diffeomorphisms f and g,

$$S(g \circ f) = S(g) \circ f + S(f) \tag{1.9}$$

where $S(g) \circ f$ is defined as in (1.7). The homological meaning of this equation will be explained in Section 1.5.

Exercise 1.3.2 Prove formula (1.9).

Curves in the projective line

By a curve we mean a parameterized curve, that is, a smooth map from $\mathbb{R}$ to $\mathbb{RP}^1$. In other words, we consider a moving one-dimensional subspace in $\mathbb{R}^2$. Two curves $\gamma_1(t)$ and $\gamma_2(t)$ are called equivalent if there exists a projective transformation $g \in \mathrm{PGL}(2, \mathbb{R})$ such that $\gamma_2(t) = g \circ \gamma_1(t)$. Recall furthermore that a curve in $\mathbb{RP}^1$ is non-degenerate if its speed is never vanishing (cf. Section 1.1).

One wants to describe the equivalence classes of non-degenerate curves in $\mathbb{RP}^1$. In answering this question we encounter, for the first time, a powerful tool of projective differential geometry: linear differential operators.

Theorem-construction 1.3.1 *There is a one-to-one correspondence between equivalence classes of non-degenerate curves in $\mathbb{RP}^1$ and Sturm–Liouville operators*

$$L = \frac{d^2}{dt^2} + u(t) \tag{1.10}$$

where $u(t)$ is a smooth function.

Proof Consider the Sturm–Liouville equation $\ddot{\psi}(t) + u(t)\psi(t) = 0$ associated with an operator (1.10). The space of solutions, V, of this equation is two-dimensional. Associating to each value of t a one-dimensional subspace of V consisting of solutions vanishing for this t, we obtain a family of one-dimensional subspaces depending on t. Finally, identifying V with $\mathbb{R}^2$ by an arbitrary choice of a basis, $\psi_1(t)$, $\psi_2(t)$, we obtain a curve in $\mathbb{RP}^1$, defined up to a projective equivalence.

Conversely, consider a non-degenerate curve $\gamma(t)$ in $\mathbb{RP}^1$. It can be uniquely lifted to $\mathbb{R}^2$ as a curve $\Gamma(t)$ such that $|\Gamma(t), \dot{\Gamma}(t)| = 1$; see Figure 1.10. Differentiate to see that the vector $\ddot{\Gamma}(t)$ is proportional to $\Gamma(t)$:

$$\ddot{\Gamma}(t) + u(t)\Gamma(t) = 0.$$

We have obtained a Sturm–Liouville operator. If $\gamma(t)$ is replaced by a projectively equivalent curve then its lift $\Gamma(t)$ is replaced by a curve $A(\Gamma(t))$ where $A \in \mathrm{SL}(2, \mathbb{R})$, and the respective Sturm–Liouville operator remains intact. □

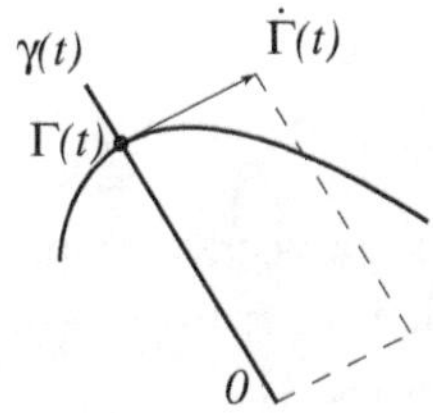

Fig. 1.10. Canonical lift of γ to $\mathbb{R}^2$: the area $|\Gamma(t), \dot{\Gamma}(t)| = 1$

Exercise 1.3.3 Show that:

(a) the curve corresponding to a Sturm–Liouville operator is non-degenerate;
(b) the two above constructions are inverse to each other.

To compute explicitly the correspondence between Sturm–Liouville operators and non-degenerate curves, fix an affine coordinate on $\mathbb{RP}^1$. A curve γ is then given by a function $f(t)$.

Exercise 1.3.4 Check that $u(t) = \frac{1}{2}S(f(t))$.

Thus the Schwarzian derivative enters the plot for the second time.

Projective structures on $\mathbb{R}$ and S^1

The definition of a projective structure resembles many familiar definitions in differential topology or differential geometry (smooth manifold, vector bundle, etc.). A projective structure on $\mathbb{R}$ is given by an atlas (U_i, φ_i) where (U_i) is an open covering of $\mathbb{R}$ and the maps $\varphi_i : U_i \to \mathbb{RP}^1$ are local diffeomorphisms satisfying the following condition: the locally defined maps $\varphi_i \circ \varphi_j^{-1}$ on $\mathbb{RP}^1$ are projective. Two such atlases are equivalent if their union is again an atlas.

Informally speaking, a projective structure is a local identification of $\mathbb{R}$ with $\mathbb{RP}^1$. For every quadruple of sufficiently close points one has the notion of cross-ratio.

A projective atlas defines an immersion $\varphi : \mathbb{R} \to \mathbb{RP}^1$; a projective structure gives a projective equivalence class of such immersions. The immersion φ, modulo projective equivalence, is called the *developing map*. According to Theorem 1.3.1, the developing map φ gives rise to a Sturm–Liouville operator (1.10). Therefore, *the space of projective structures on S^1 is identified with the space of Sturm–Liouville operators.*

The definition of a projective structure on S^1 is analogous, but it has a new feature. Identifying S^1 with $\mathbb{R}/\mathbb{Z}$, the developing map satisfies the following condition:

$$\varphi(t+1) = M(\varphi(t)) \tag{1.11}$$

for some $M \in \mathrm{PGL}(2, \mathbb{R})$. The projective map M is called the *monodromy*. Again, the developing map is defined up to the projective equivalence: $(\varphi(t), M) \sim (g\varphi(t), gMg^{-1})$ for $g \in \mathrm{PGL}(2, \mathbb{R})$.

The monodromy condition (1.11) implies that, for the corresponding Sturm–Liouville operator, one has $u(t+1) = u(t)$, while the solutions have the monodromy $\widetilde{M} \in \mathrm{SL}(2, \mathbb{R})$, which is a lift of M. To summarize, the space of projective structures on S^1 is identified with the space of Sturm–Liouville operators with 1-periodic potentials $u(t)$.

$\mathrm{Diff}(S^1)$- and $\mathrm{Vect}(S^1)$-action on projective structures

The group of diffeomorphisms $\mathrm{Diff}(S^1)$ naturally acts on projective atlases and, therefore, on the space of projective structures. In terms of the Sturm–Liouville operators, this action is given by the transformation rule for the potential

$$\mathrm{T}_{f^{-1}} : u \mapsto \left(f'\right)^2 u(f) + \tfrac{1}{2}\, S(f), \tag{1.12}$$

where $f \in \mathrm{Diff}(S^1)$. This follows from Exercise 1.3.4 and formula (1.9).

The Lie algebra corresponding to $\mathrm{Diff}(S^1)$ is the algebra of vector fields $\mathrm{Vect}(S^1)$. The vector fields are written as $X = h(t)\mathrm{d}/\mathrm{d}t$ and their commutator as

$$[X_1, X_2] = \left(h_1 h_2' - h_1' h_2\right) \frac{\mathrm{d}}{\mathrm{d}t}.$$

Whenever one has a differentiable action of $\mathrm{Diff}(S^1)$, one also has an action of $\mathrm{Vect}(S^1)$ on the same space.

Exercise 1.3.5 Check that the action of a vector field $X = h(t)\mathrm{d}/\mathrm{d}t$ on the potential of a Sturm–Liouville operator is given by

$$\mathrm{t}_X : u \mapsto hu' + 2h'u + \tfrac{1}{2}\, h'''. \tag{1.13}$$

It is interesting to describe the kernel of this action.

Exercise 1.3.6 (a) Let ϕ_1 and ϕ_2 be two solutions of the Sturm–Liouville equation $\phi''(t) + u(t)\phi(t) = 0$. Check that, for the vector field $X = \phi_1\phi_2\,\mathrm{d}/\mathrm{d}t$, one has $\mathrm{t}_X = 0$.
(b) Show that the kernel of the action t is a Lie algebra isomorphic to $\mathrm{sl}(2, \mathbb{R})$; this is precisely the Lie algebra of symmetries of the projective structure corresponding to the Sturm–Liouville operator.

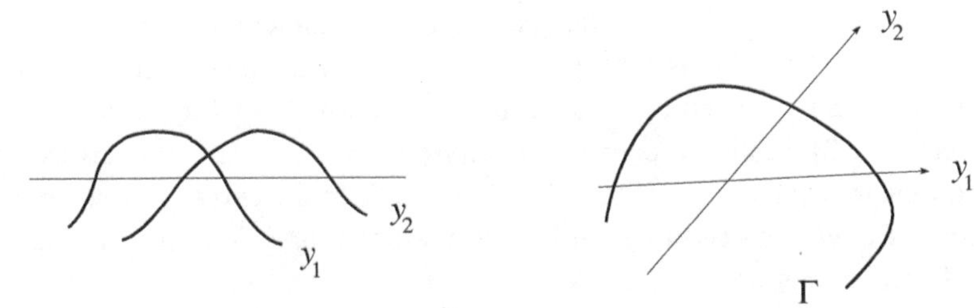

Fig. 1.11. Zeroes of solutions

Hint. The space of solutions of the equation $t_X = 0$ is three-dimensional, hence the products of two solutions of the Sturm–Liouville equation span this space.

Sturm theorem on zeroes

The classic Sturm theorem states that *between two zeroes of a solution of a Sturm–Liouville equation any other solution has a zero as well.* The simplest proof is an application of the above identification between Sturm–Liouville equations and projective structures on S^1. Consider the corresponding developing map $\gamma : S^1 \to \mathbb{RP}^1$ and its lift Γ to $\mathbb{R}^2$. Every solution ϕ of the Sturm–Liouville equation is a pull-back of a linear function y on $\mathbb{R}^2$. Zeroes of ϕ are the intersection points of Γ with the line $y = 0$. Since γ is non-degenerate, the intermediate value theorem implies that between two intersections of Γ with any line there is an intersection with any other line; see Figure 1.11 and [163] for an elementary exposition.

Comment

The Schwarzian derivative is historically the first and most fundamental projective differential invariant. The natural identification of the space of projective structures with the space of Sturm–Liouville operators is an important conceptual result of one-dimensional projective differential geometry; see [222] for a survey. Exercise 1.3.6 is Kirillov's observation [115].

1.4 A further example of differential invariants: projective curvature

The second oldest differential invariant of projective geometry is the projective curvature of a plane curve. The term "curvature" is somewhat misleading: the projective curvature is by no means a function on the curve. We will define the projective curvature as a projective structure on the curve. In a nutshell, the curve is approximated by its osculating conic which, by Steiner's theorem (cf. Section 1.2), has a projective structure induced from $\mathbb{RP}^2$; this projective structure is transplanted from the osculating conic to the curve. To realize this program, we will proceed in a traditional way and represent projective curves by differential operators.

Plane curves and differential operators

Consider a parameterized non-degenerate curve $\gamma(t)$ in $\mathbb{RP}^2$, that is, a curve without inflection points (see Section 1.1 for a general definition). Repeating the construction of Theorem 1.3.1 yields a third-order linear differential operator

$$A = \frac{\mathrm{d}^3}{\mathrm{d}t^3} + q(t)\frac{\mathrm{d}}{\mathrm{d}t} + r(t). \tag{1.14}$$

Example 1.4.1 Let $\gamma(t)$ be the conic (recall that all non-degenerate conics in $\mathbb{RP}^2$ are projectively equivalent). The corresponding differential operator (1.14) has a special form:

$$A_1 = \frac{\mathrm{d}^3}{\mathrm{d}t^3} + q(t)\frac{\mathrm{d}}{\mathrm{d}t} + \frac{1}{2}q'(t). \tag{1.15}$$

Indeed, consider the *Veronese map* $V : \mathbb{RP}^1 \to \mathbb{RP}^2$ given by the formula

$$V(x_0 : x_1) = (x_0^2 : x_0x_1 : x_1^2). \tag{1.16}$$

The image of $\mathbb{RP}^1$ is a conic, and $\gamma(t)$ is the image of a parameterized curve in $\mathbb{RP}^1$. A parameterized curve in $\mathbb{RP}^1$ corresponds to a Sturm–Liouville operator (1.10) so that $\{x_0(t), x_1(t)\}$ is a basis of solutions of the Sturm–Liouville equation $L\psi = 0$. It remains to check that every product

$$y(t) = x_i(t)x_j(t), \qquad i, j = 1, 2,$$

satisfies $A_1 y = 0$ with $q(t) = 4u(t)$.

Exercise 1.4.2 We now have two projective structures on the conic in $\mathbb{RP}^2$: the one given by Steiner's theorem and the one induced by the Veronese map from $\mathbb{RP}^1$. Prove that these structures coincide.

Projective curvature via differential operators

Associate the following Sturm–Liouville operator with the operator A:

$$L = \frac{\mathrm{d}^2}{\mathrm{d}t^2} + \frac{1}{4} q(t). \tag{1.17}$$

According to Section 1.3, we obtain a projective structure on $\mathbb{R}$ and thus on the parameterized curve $\gamma(t)$.

Theorem 1.4.3 *This projective structure on $\gamma(t)$ does not depend on the choice of the parameter t.*

Proof Recall the notion of dual (or adjoint) operator: for a differential monomial one has

$$\left(a(t)\frac{\mathrm{d}^k}{\mathrm{d}t^k}\right)^* = (-1)^k \frac{\mathrm{d}^k}{\mathrm{d}t^k} \circ a(t). \tag{1.18}$$

Consider the decomposition of the operator (1.14) into the sum

$$A = A_1 + A_0 \tag{1.19}$$

of its skew-symmetric part $A_1 = -A_1^*$ given by (1.15) and the symmetric part $A_0 = A_0^*$. Note, that the symmetric part is a scalar operator:

$$A_0 = r(t) - \tfrac{1}{2} q'(t). \tag{1.20}$$

The decomposition (1.19) is *intrinsic*, that is, independent of the choice of the parameter t (cf. Section 2.2 below).

The correspondence $A \mapsto L$ is a composition of two operations: $A \mapsto A_1$ and $A_1 \mapsto L$; the second one is also intrinsic, cf. Example 1.4.1. □

Exercise 1.4.4 Show that the operator (1.15) is skew-symmetric: $A_1^* = -A_1$.

To wit, a non-degenerate curve in $\mathbb{RP}^2$ carries a canonical projective structure which we call the *projective curvature*. In the next chapter we will explain that the expression A_0 in (1.20) is, in fact, a cubic differential; the cube root $(A_0)^{1/3}$ is called the *projective length element*. The projective length element is identically zero for a conic and, moreover, vanishes in those points of the curve in which the osculating conic is hyper-osculating.

Traditionally, the projective curvature is considered as a function $q(t)$ where t is a special parameter for which $A_0 \equiv 1$, that is, the projective length element equals $\mathrm{d}t$.

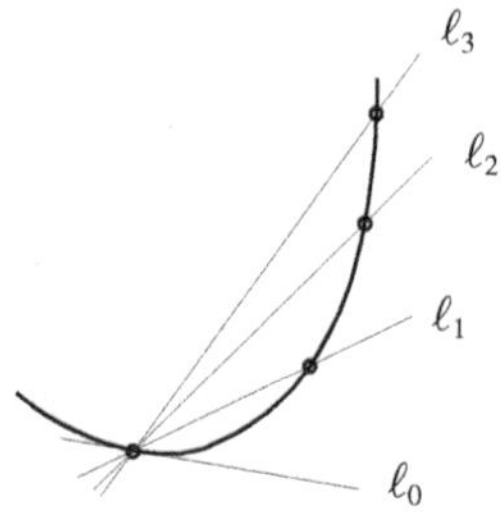

Fig. 1.12. Projective curvature as cross-ratio

On the other hand, one can choose a different parameter x on the curve in such a way that $q(x) \equiv 0$, namely, the affine coordinate of the defined projective structure. This shows that the projective curvature is neither a function nor a tensor.

Exercise 1.4.5 (a) Let A be the differential operator corresponding to a non-degenerate parameterized curve $\gamma(t)$ in $\mathbb{RP}^2$. Prove that the operator corresponding to the dual curve $\gamma^*(t)$ is $-A^*$.
(b) Consider a non-degenerate parameterized curve $\gamma(t)$ in $\mathbb{RP}^2$ and let $\gamma^*(t)$ be projectively equivalent to $\gamma(t)$, that is, there exists a projective isomorphism $\varphi : \mathbb{RP}^2 \to \mathbb{RP}^{2*}$ such that $\gamma^*(t) = \varphi(\gamma(t))$. Prove that $\gamma(t)$ is a conic.

Projective curvature and cross-ratio

Consider four points

$$\gamma(t), \quad \gamma(t+\varepsilon), \quad \gamma(t+2\varepsilon), \quad \gamma(t+3\varepsilon)$$

of a non-degenerate curve in $\mathbb{RP}^2$. These points determine four lines ℓ_0, ℓ_1, ℓ_2 and ℓ_3 as in Figure 1.12.

Let us expand the cross-ratio of these lines in powers of ε.

Exercise 1.4.6 Show that

$$[\ell_0, \ell_1, \ell_2, \ell_3] = 4 - 2\varepsilon^2 q(t) + O(\varepsilon^3). \tag{1.21}$$

This formula relates the projective curvature with the cross-ratio.

Comparison with affine curvature

Let us illustrate the preceding construction by comparison with geometrically more transparent notion of the affine curvature and the affine parameter (see, e.g., [193]).

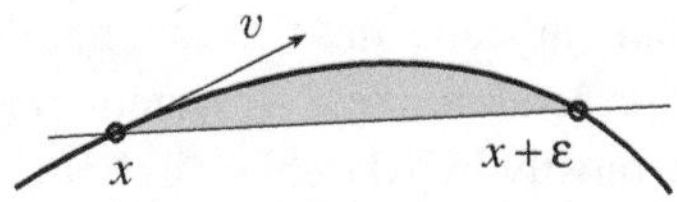

Fig. 1.13. Cubic form on an affine curve

Consider a non-degenerate curve γ in the affine plane with a fixed area form. We define a cubic form on γ as follows. Let v be a tangent vector to γ at point x. Extend v to a tangent vector field along γ and denote by ϕ_t the corresponding local one-parameter group of diffeomorphisms of γ. Consider the segment between x and $\phi_\varepsilon(x)$ (see Figure 1.13), and denote by $A(x, v, \varepsilon)$ the area bounded by it and the curve. This function behaves cubically in ε, and we define a cubic form

$$\sigma(x, v) = \lim_{\varepsilon\to 0} \frac{A(x, v, \varepsilon)}{\varepsilon^3}. \tag{1.22}$$

A parameter t on γ is called affine if $\sigma = c(\mathrm{d}t)^3$ with a positive constant c. By the very construction, the notion of affine parameter is invariant with respect to the group of affine transformations of the plane while σ is invariant under the (smaller) equiaffine group.

Alternatively, an affine parameter is characterized by the condition

$$|\gamma'(t), \gamma''(t)| = \text{const}.$$

Hence the vectors $\gamma'''(t)$ and $\gamma'(t)$ are proportional: $\gamma'''(t) = -k(t)\gamma'(t)$. The function $k(t)$ is called the *affine curvature*.

The affine parameter is not defined at inflection points. The affine curvature is constant if and only if γ is a conic.

Comment

The notion of projective curvature appeared in the literature in the second half of the nineteenth century. From the very beginning, curves were studied in the framework of differential operators – see [231] for an account of this early period of projective differential geometry.

In his book [37], E. Cartan also calculated the projective curvature as a function of the projective length parameter. However, he gave an interpretation of the projective curvature in terms of a projective structure on the curve. Cartan invented a geometrical construction of developing a non-degenerate curve on its osculating conic. This construction is a projective counterpart of the Huygens construction of the involute of a plane curve using a non-stretchable string: the role of the tangent line is played by the osculating conic and the role of the Euclidean length by the projective one.

Affine differential geometry and the corresponding differential invariants appeared later than the projective ones. A systematic theory was developed between 1910 and 1930, mostly by Blaschke's school.

1.5 The Schwarzian derivative as a cocycle of $\mathrm{Diff}(\mathbb{RP}^1)$

The oldest differential invariant of projective geometry, the Schwarzian derivative, remains the most interesting one. In this section we switch gears and discuss the relation of the Schwarzian derivative with cohomology of the group $\mathrm{Diff}(\mathbb{RP}^1)$. This contemporary viewpoint leads to promising applications that will be discussed later in the book. To better understand the material of this and the next section, the reader is recommended to consult Appendix A.4.

Invariant and relative 1-cocycles

Let G be a group, V a G-module and $T : G \to \mathrm{End}(V)$ the G-action on V. A map $C : G \to V$ is called a 1-cocycle on G with coefficients in V if it satisfies the condition

$$C(gh) = T_g\, C(h) + C(g). \tag{1.23}$$

A 1-cocycle C is called a coboundary if

$$C(g) = T_g\, v - v \tag{1.24}$$

for some fixed $v \in V$. The quotient group of 1-cocycles by coboundaries is $H^1(G, V)$, the first cohomology group; see Appendix A.4 for more details.

Let H be a subgroup of G. A 1-cocycle C is H-*invariant* if

$$C(hgh^{-1}) = T_h\, C(g) \tag{1.25}$$

for all $h \in H$ and $g \in G$.

Another important class of 1-cocycles associated with a subgroup H consists of the cocycles vanishing on H. Such cocycles are called H-*relative*.

Exercise 1.5.1 Let H be a subgroup of G and let C be a 1-cocycle on G. Prove that the following three conditions are equivalent:

(1) $C(h) = 0$ for all $h \in H$;
(2) $C(gh) = C(g)$ for all $h \in H$ and $g \in G$;
(3) $C(hg) = T_h\,(C(g))$ for all $h \in H$ and $g \in G$.

The property of a 1-cocycle to be H-relative is stronger than the condition to be H-invariant.

Exercise 1.5.2 Check that the conditions (1)–(3) imply (1.25).

Tensor densities in dimension 1

All tensor fields on a one-dimensional manifold M are of the form:

$$\phi = \phi(x)(\mathrm{d}x)^\lambda, \tag{1.26}$$

where $\lambda \in \mathbb{R}$ and x is a local coordinate; ϕ is called a *tensor density* of degree λ. The space of tensor densities is denoted by $\mathcal{F}_\lambda(M)$, or $\mathcal{F}_\lambda$ for short. Equivalently, a tensor density of degree λ is defined as a section of the line bundle $(T^*M)^{\otimes\lambda}$.

The group Diff(M) naturally acts on $\mathcal{F}_\lambda$. To describe explicitly this action, consider the space of functions $C^\infty(M)$ and define a one-parameter family of Diff(M)-actions on this space:

$$\mathrm{T}^\lambda_{f^{-1}} : \phi(x) \mapsto \left(f'\right)^\lambda \phi(f(x)), \qquad f \in \mathrm{Diff}(M) \tag{1.27}$$

cf. formula (1.7) for quadratic differentials. The Diff(M)-module $\mathcal{F}_\lambda$ is nothing else but the module $(C^\infty(M), \mathrm{T}^\lambda)$. Although all $\mathcal{F}_\lambda$ are isomorphic to each other as vector spaces, $\mathcal{F}_\lambda$ and $\mathcal{F}_\mu$ are not isomorphic as Diff(M)-modules unless $\lambda = \mu$ (cf. [72]).

In the case $M = S^1$, there is a Diff(M)-invariant pairing $\mathcal{F}_\lambda \otimes \mathcal{F}_{1-\lambda} \to \mathbb{R}$ given by the integral

$$\langle \phi(x)(\mathrm{d}x)^\lambda, \psi(x)(\mathrm{d}x)^{1-\lambda} \rangle = \int_{S^1} \phi(x)\psi(x)\mathrm{d}x$$

Example 1.5.3 In particular, $\mathcal{F}_0$ is the space of smooth functions, $\mathcal{F}_1$ is the space of 1-forms, $\mathcal{F}_2$ is the space of quadratic differentials, familiar from the definition of the Schwarzian derivative, while $\mathcal{F}_{-1}$ is the space of vector fields. The whole family $\mathcal{F}_\lambda$ is of importance, especially for integer and half-integer values of λ.

Exercise 1.5.4 (a) Check that formula (1.27) indeed defines an action of Diff(M), that is, for all diffeomorphisms f, g, one has $\mathrm{T}^\lambda_f \circ \mathrm{T}^\lambda_g = \mathrm{T}^\lambda_{f \circ g}$.
(b) Show that the Vect(M)-action on $\mathcal{F}_\lambda(M)$ is given by the formula

$$L^\lambda_{h(x)\frac{\mathrm{d}}{\mathrm{d}x}} : \phi(\mathrm{d}x)^\lambda \mapsto (h\phi' + \lambda\, h'\phi)(\mathrm{d}x)^\lambda. \tag{1.28}$$

First cohomology with coefficients in tensor densities

Recall identity (1.9) for the Schwarzian derivative. This identity means that the Schwarzian derivative defines a 1-cocycle $f \mapsto S(f^{-1})$ on $\mathrm{Diff}(\mathbb{RP}^1)$ with coefficients in the space of quadratic differentials $\mathcal{F}_2(\mathbb{RP}^1)$. This cocycle is not a coboundary; indeed, unlike $S(f)$, any coboundary (1.24) depends only on the 1-jet of a diffeomorphism – see formula (1.27).

The Schwarzian derivative vanishes on the subgroup $\mathrm{PGL}(2, \mathbb{R})$, and thus it is $\mathrm{PGL}(2, \mathbb{R})$-invariant.

Let us describe the first cohomology of the group $\mathrm{Diff}(\mathbb{RP}^1)$ with coefficients in $\mathcal{F}_\lambda$. These cohomologies can be interpreted as equivalence classes of affine modules (or extensions) on $\mathcal{F}_\lambda$. If G is a Lie group and V its module then a structure of an affine module on V is a structure of a G-module on the space $V \oplus \mathbb{R}$ defined by

$$\widetilde{T}_g : (v, \alpha) \mapsto (T_g\, v + \alpha\, C(g), \alpha),$$

where C is a 1-cocycle on G with values in V. See Appendix A.4 for more information on affine modules and extensions.

Theorem 1.5.5 *One has*

$$H^1(\mathrm{Diff}(\mathbb{RP}^1); \mathcal{F}_\lambda) = \begin{cases} \mathbb{R}, & \lambda = 0, 1, 2, \\ 0, & \text{otherwise.} \end{cases} \tag{1.29}$$

We refer to [72] for details. The corresponding cohomology classes are represented by the 1-cocycles

$$C_0(f^{-1}) = \ln f', \quad C_1(f^{-1}) = \frac{f''}{f'}\, \mathrm{d}x, \quad C_2(f^{-1}) = \left(\frac{f'''}{f'} - \frac{3}{2}\left(\frac{f''}{f'}\right)^2 \right) (\mathrm{d}x)^2.$$

The first cocycle makes sense in Euclidean geometry and the second one in affine geometry. Their restrictions to the subgroup $\mathrm{PGL}(2, \mathbb{R})$ are non-trivial, hence these cohomology classes cannot be represented by $\mathrm{PGL}(2, \mathbb{R})$-relative cocycles.

One is usually interested in cohomology classes, not in the representing cocycles which, as a rule, depend on arbitrary choices. However, the Schwarzian derivative is canonical in the following sense.

Theorem 1.5.6 *The Schwarzian derivative is a unique (up to a constant)* $\mathrm{PGL}(2, \mathbb{R})$*-relative 1-cocycle on* $\mathrm{Diff}(\mathbb{RP}^1)$ *with coefficients in* $\mathcal{F}_2$.

Proof If there are two such cocycles, then, by Theorem 1.5.5, their linear combination is a coboundary, and this coboundary vanishes on $\mathrm{PGL}(2, \mathbb{R})$. Every coboundary is of the form

$$C(f) = \mathrm{T}_f(\phi) - \phi$$

for some $\phi \in \mathcal{F}_2$. Therefore one has a non-zero $\mathrm{PGL}(2, \mathbb{R})$-invariant quadratic differential. It remains to note that $\mathrm{PGL}(2, \mathbb{R})$ does not preserve any tensor field on $\mathbb{RP}^1$. □

Exercise 1.5.7 Prove that the infinitesimal version of the Schwarzian derivative is the following 1-cocycle on the Lie algebra $\mathrm{Vect}(\mathbb{RP}^1)$:

$$h(x)\frac{\mathrm{d}}{\mathrm{d}x} \mapsto h'''(x)\,(\mathrm{d}x)^2. \tag{1.30}$$

1.6 Virasoro algebra: the coadjoint representation

The Virasoro algebra is one of the best-known infinite-dimensional Lie algebras, defined as a central extension of $\mathrm{Vect}(S^1)$. A central extension of a Lie algebra $\mathfrak{g}$ is a Lie algebra structure on the space $\mathfrak{g} \oplus \mathbb{R}$ given by the commutator

$$[(X, \alpha), (Y, \beta)] = ([X, Y], c(X, Y)),$$

where $X, Y \in \mathfrak{g}$, $\alpha, \beta \in \mathbb{R}$ and $c : \mathfrak{g} \to \mathbb{R}$ is a 1-cocycle. The reader can find more information on central extensions in Appendix A.4.

Definition of the Virasoro algebra

The Lie algebra $\mathrm{Vect}(S^1)$ has a central extension given by the so-called Gelfand–Fuchs cocycle

$$c\left(h_1(x)\frac{\mathrm{d}}{\mathrm{d}x}, h_2(x)\frac{\mathrm{d}}{\mathrm{d}x}\right) = \int_{S^1} h_1'(x)\,h_2''(x)\,\mathrm{d}x. \tag{1.31}$$

The corresponding Lie algebra is called the Virasoro algebra and will be denoted by Vir. This is a unique (up to isomorphism) non-trivial central extension of $\mathrm{Vect}(S^1)$ (cf. Lemma A.5.3).

Exercise 1.6.1 Check the Jacobi identity for Vir.

Note that the cocycle (1.31) is obtained by pairing the cocycle (1.30) with a vector field.

Computing the coadjoint representation

To explain the relation of the Virasoro algebra to projective geometry we use the notion of *coadjoint representation* defined as follows. A Lie algebra $\mathfrak{g}$ acts on its dual space by

$$\langle \mathrm{ad}^*_X \phi, Y \rangle := -\langle \phi, [X, Y] \rangle,$$

for $\phi \in \mathfrak{g}^*$ and $X, Y \in \mathfrak{g}$. This coadjoint representation carries much information about the Lie algebra.

The dual space to the Virasoro algebra is $\mathrm{Vir}^* = \mathrm{Vect}(S^1)^* \oplus \mathbb{R}$. It is always natural to begin the study of the dual space to a functional space with its subspace called the *regular dual*. This subspace is spanned by the distributions given by smooth compactly supported functions.

Consider the regular dual space, $\mathrm{Vir}^*_{\mathrm{reg}} = C^\infty(S^1) \oplus \mathbb{R}$, consisting of pairs $(u(x), c)$ where $u(x) \in C^\infty(S^1)$ and $c \in \mathbb{R}$, so that

$$\langle (u(x), c), (h(x)\mathrm{d}/\mathrm{d}x, \alpha) \rangle := \int_{S^1} u(x)h(x)\mathrm{d}x + c\alpha.$$

The regular dual space is invariant under the coadjoint action.

Exercise 1.6.2 Show that the explicit formula for the coadjoint action of the Virasoro algebra on its regular dual space is

$$\mathrm{ad}^*_{(h\mathrm{d}/\mathrm{d}x,\, \alpha)}(u, c) = (hu' + 2h'u - c\, h''', 0). \tag{1.32}$$

Note that the center of Vir acts trivially.

A remarkable coincidence

In the first two terms of the above formula (1.32) we recognize the Lie derivative (1.28) of quadratic differentials, the third term is nothing else but the cocycle (1.30), so that the action (1.32) is an affine module (see Section 1.5). Moreover, this action coincides with the natural $\mathrm{Vect}(S^1)$-action on the space of Sturm–Liouville operators (for $c = -1/2$); see formula (1.13).

Thus one identifies as $\mathrm{Vect}(S^1)$-modules the regular dual space $\mathrm{Vir}^*_{\mathrm{reg}}$ and the space of Sturm–Liouville operators

$$(u(x), c) \leftrightarrow -2c \frac{\mathrm{d}^2}{\mathrm{d}x^2} + u(x) \tag{1.33}$$

and obtains a nice geometrical interpretation for the coadjoint representation of the Virasoro algebra.

Remark 1.6.3 To simplify exposition, we omit the definition of the Virasoro group (the group analog of the Virasoro algebra) and the computation of its coadjoint action which, indeed, coincides with the $\mathrm{Diff}(S^1)$-action (1.12).

Coadjoint orbits

The celebrated Kirillov's orbit method concerns the study of the coadjoint representation. Coadjoint orbits of a Lie algebra $\mathfrak{g}$ are defined as integral surfaces in $\mathfrak{g}^*$, tangent to the vector fields $\dot{\phi} = \mathrm{ad}^*_X \phi$ for all $X \in \mathfrak{g}$.[2] Classification of the coadjoint orbits of a Lie group or a Lie algebra is always an interesting problem.

The identification (1.33) makes it possible to express invariants of the coadjoint orbits of the Virasoro algebra in terms of invariants of Sturm–Liouville operators (and projective structures on S^1, see Theorem 1.3.1).

An invariant of a differential operator on S^1 is the monodromy operator mentioned in Section 1.3. In the case of Sturm–Liouville operators, this is an element of the universal covering $\widetilde{\mathrm{PGL}(2, \mathbb{R})}$.

Theorem 1.6.4 *The monodromy operator is the unique invariant of the coadjoint orbits of the Virasoro algebra.*

Proof Two elements $(u_0(x), c)$ and $(u_1(x), c)$ of $\mathrm{Vir}^*_{\mathrm{reg}}$ belong to the same coadjoint orbit if and only if there is a one-parameter family $(u_t(x), c)$ with $t \in [0, 1]$ such that, for every t, the element $(\dot{u}_t(x), 0)$ is the result of the coadjoint action of Vir; here dot denotes the derivative with respect to t. In other words, there exists $h_t(x)\frac{\mathrm{d}}{\mathrm{d}x} \in \mathrm{Vect}(S^1)$ such that

$$\dot{u}_t(x) = h_t(x)\, u_t'(x) + 2h_t'(x)\, u_t(x) - c\, h_t'''(x). \tag{1.34}$$

According to (1.33), a family $(u_t(x), c)$ defines a family of Sturm–Liouville operators: $L_t = -2c(\mathrm{d}/\mathrm{d}x)^2 + u(x)_t$. Consider the corresponding family of Sturm–Liouville equations

$$L_t(\phi) = -2c\, \phi''(x) + u(x)_t\, \phi(x) = 0.$$

For every t, one has a two-dimensional space of solutions, $\langle \phi_{1_t}(x), \phi_{2_t}(x) \rangle$.

Define a $\mathrm{Vect}(S^1)$-action on the space of solutions using the Leibnitz rule:

$$(\mathrm{ad}^*_{h\mathrm{d}/\mathrm{d}x} L)(\phi) + L(T_{h\mathrm{d}/\mathrm{d}x}\phi) = 0$$

[2] This definition allows us to avoid using the notion of a Lie group, and sometimes this simplifies the situation, for instance, in the infinite-dimensional case.

where the Vect(S^1)-action on the space of Sturm–Liouville operators is given by formula (1.32). It turns out that the solutions of Sturm–Liouville equations behave as tensor densities of degree $-\frac{1}{2}$.

Exercise 1.6.5 Check that, in the above formula, $T_{h\mathrm{d}/\mathrm{d}x} = L^{-1/2}_{h\mathrm{d}/\mathrm{d}x}$, where $L^{\lambda}_{h\mathrm{d}/\mathrm{d}x}$ is the Lie derivative of a λ-density defined by (1.28).

To solve the (nonlinear) "homotopy" equation (1.34), it suffices now to find a family of vector fields $h_t(x)\mathrm{d}/\mathrm{d}x$ such that

$$\begin{cases} L^{-1/2}_{h_t\mathrm{d}/\mathrm{d}x}\,\phi_{1_t} = h_t\,\phi_{1_t}' - \frac{1}{2}\,h_t'\,\phi_{1_t} = \dot{\phi}_{1_t}, \\ L^{-1/2}_{h_t\mathrm{d}/\mathrm{d}x}\,\phi_{2_t} = h_t\,\phi_{2_t}' - \frac{1}{2}\,h_t'\,\phi_{2_t} = \dot{\phi}_{2_t}. \end{cases}$$

This is just a system of linear equations in two variables, $h_t(x)$ and $h_t'(x)$, with the solution

$$h_t(x) = \begin{vmatrix} \dot{\phi}_{1_t} & \dot{\phi}_{2_t} \\ \phi_{1_t} & \phi_{2_t} \end{vmatrix}, \qquad h_t'(x) = 2\begin{vmatrix} \dot{\phi}_{1_t} & \dot{\phi}_{2_t} \\ \phi_{1_t}' & \phi_{2_t}' \end{vmatrix}. \tag{1.35}$$

One can choose a basis $\langle \phi_{1_t}(x), \phi_{2_t}(x)\rangle$ so that the Wronski determinant is independent of t:

$$\begin{vmatrix} \phi_{1_t} & \phi_{2_t} \\ \phi_{1_t}' & \phi_{2_t}' \end{vmatrix} \equiv 1.$$

Then one has

$$\begin{vmatrix} \dot{\phi}_{1_t} & \dot{\phi}_{2_t} \\ \phi_{1_t}' & \phi_{2_t}' \end{vmatrix} = \begin{vmatrix} \dot{\phi}_{1_t}' & \dot{\phi}_{2_t}' \\ \phi_{1_t} & \phi_{2_t} \end{vmatrix}$$

It follows that the first formula in (1.35) implies the second.

Finally, if the monodromy operator of a family of Sturm–Liouville operators L_t does not depend on t, then one can choose a basis $\langle \phi_{1_t}(x), \phi_{2_t}(x)\rangle$ in such a way that the monodromy matrix, say M, in this basis does not depend on t. Then one concludes from (1.35) that

$$h_t(x + 2\pi) = \det M \cdot h_t(x) = h_t(x),$$

since $M \in \widetilde{\mathrm{SL}(2,\mathbb{R})}$. Therefore, $h_t(x)\mathrm{d}/\mathrm{d}x$ is, indeed, a family of vector fields on S^1. □

Remark 1.6.6 One can understand, in a more traditional way, the monodromy operator as an element of $\mathrm{SL}(2,\mathbb{R})$, instead of its universal covering. Then there is another discrete invariant, representing a class in $\pi_1(\mathrm{SL}(2,\mathbb{R}))$. This invariant is nothing but the winding number of the corresponding curve in $\mathbb{RP}^1$;

see Section 1.3. For instance, there are infinitely many connected components in the space of Sturm–Liouville operators with the same monodromy.

Relation to infinite-dimensional symplectic geometry

A fundamental fact which makes the notion of coadjoint orbits so important (in comparison with the adjoint orbits) is that every coadjoint orbit has a canonical $\mathfrak{g}$-invariant symplectic structure (often called the Kirillov symplectic form). Moreover, the space $\mathfrak{g}^*$ has a Poisson structure called the Lie–Poisson (–Berezin–Kirillov–Kostant–Souriau) bracket, and the coadjoint orbits are the corresponding symplectic leaves. See Appendix A.2 for a brief introduction to symplectic and Poisson geometry.

An immediate corollary of the above remarkable coincidence is that the space of the Sturm–Liouville operators is endowed with a natural $\mathrm{Diff}(S^1)$-invariant Poisson structure; furthermore, it follows from Theorem 1.6.4 that the space of Sturm–Liouville operators with a fixed monodromy is an (infinite-dimensional) symplectic manifold.

Comment

The Virasoro algebra was discovered in 1967 by I. M. Gelfand and D. B. Fuchs. It appeared in the physical literature around 1975 and became very popular in conformal field theory (see [90] for a comprehensive reference).

The coadjoint representation of Lie groups and Lie algebras plays a special role in symplectic geometry and representation theory, cf. [112]. The observation relating the coadjoint representation to the natural $\mathrm{Vect}(S^1)$-action on the space of Sturm–Liouville operators and, therefore, on the space of projective structures on S^1, and the classification of the coadjoint orbits was made in 1980 independently by A. A. Kirillov and G. Segal [116, 186]. The classification of the coadjoint orbits then follows from the classical work by Kuiper [126] (see also [130]) on classification of projective structures. Our proof, using the homotopy method, is probably new.

This and other remarkable properties of the Virasoro algebra, its relation with the Korteweg–de Vries equation, moduli spaces of holomorphic curves, etc., make this infinite-dimensional Lie algebra one of the most interesting objects of modern mathematics and mathematical physics.

2
The geometry of the projective line

What are geometric objects? On the one hand, curves, surfaces, various geometric structures; on the other, tensor fields, differential operators, Lie group actions. The former objects originated in classical geometry while the latter ones are associated with algebra. Both points of view are legitimate, yet often separated.

This chapter illustrates unity of geometric and algebraic approaches. We study geometry of a simple object, the projective line. Such notions as non-degenerate immersions of a line in projective space and linear differential operators on the line are intrinsically related, and this gives two complementary viewpoints on the same thing.

Following F. Klein, we understand geometry in terms of group actions. In the case of the projective line, two groups play prominent roles: the group $\mathrm{PGL}(2, \mathbb{R})$ of projective symmetries and the infinite-dimensional full group of diffeomorphisms $\mathrm{Diff}(\mathbb{RP}^1)$. We shall see how these two types of symmetry interact.

2.1 Invariant differential operators on $\mathbb{RP}^1$

The language of invariant differential operators is an adequate language for differential geometry. The best-known invariant differential operators are the de Rham differential of differential forms and the commutator of vector fields. These operators are invariant with respect to the action of the group of diffeomorphisms of the manifold. The expressions that describe these operations are independent of the choice of local coordinates.

If a manifold M carries a geometric structure, the notion of the invariant differential operator changes accordingly: the full group of diffeomorphisms is restricted to the groups preserving the geometric structure. For instance, on a symplectic manifold M, one has the Poisson bracket, a binary invariant operation on the space of smooth functions, as well as the unary operation assigning the Hamiltonian vector field to a smooth function. Another example, known to every student of calculus, is the divergence: the operator on a manifold with a fixed volume form assigning the function $\mathrm{Div} X$ to a vector field X. This operator is invariant with respect to volume preserving diffeomorphisms.

The space of differential operators $\mathcal{D}_{\lambda,\mu}(S^1)$

Consider the space of linear differential operators on S^1 from the space of λ-densities to the space of μ-densities

$$A : \mathcal{F}_\lambda(S^1) \to \mathcal{F}_\mu(S^1)$$

with arbitrary $\lambda, \mu \in \mathbb{R}$. This space will be denoted by $\mathcal{D}_{\lambda,\mu}(S^1)$ and its subspace of operators of order $\leq k$ by $\mathcal{D}^k_{\lambda,\mu}(S^1)$.

The space $\mathcal{D}_{\lambda,\mu}(S^1)$ is acted upon by $\mathrm{Diff}(S^1)$; this action is as follows:

$$\mathrm{T}^{\lambda,\mu}_f(A) = \mathrm{T}^\mu_f \circ A \circ \mathrm{T}^\lambda_{f^{-1}}, \qquad f \in \mathrm{Diff}(S^1) \tag{2.1}$$

where T^λ is the $\mathrm{Diff}(S^1)$-action on tensor densities (1.27).

For any parameter x on S^1, a k-th order differential operator is of the form

$$A = a_k(x)\frac{\mathrm{d}^k}{\mathrm{d}x^k} + a_{k-1}(x)\frac{\mathrm{d}^{k-1}}{\mathrm{d}x^{k-1}} + \cdots + a_0(x),$$

where $a_i(x)$ are smooth functions on S^1.

Exercise 2.1.1 Check that the expression

$$\sigma(A) = a_k(x)(\mathrm{d}x)^{\mu-\lambda-k}$$

does not depend on the choice of the parameter.

The density $\sigma(A)$ is called the principal symbol of A; it is a well-defined tensor density of degree $\mu - \lambda - k$. The principal symbol provides a $\mathrm{Diff}(S^1)$-invariant projection

$$\sigma : \mathcal{D}^k_{\lambda,\mu}(S^1) \to \mathcal{F}_{\mu-\lambda-k}(S^1). \tag{2.2}$$

Linear projectively invariant operators

Our goal is to describe differential operators on $\mathbb{RP}^1$, invariant under projective transformations. In the one-dimensional case, there is only one type of tensors, namely tensor densities $\phi(x)(dx)^\lambda$. Recall that the space of such tensor densities is denoted by $\mathcal{F}_\lambda(\mathbb{RP}^1)$.

A classical result of projective differential geometry is classification of projectively invariant linear differential operators $A : \mathcal{F}_\lambda(\mathbb{RP}^1) \to \mathcal{F}_\mu(\mathbb{RP}^1)$ (see [28]).

Theorem 2.1.2 *The space of* PGL(2, $\mathbb{R}$)*-invariant linear differential operators on tensor densities is generated by the identity operator from* $\mathcal{F}_\lambda(\mathbb{RP}^1)$ *to* $\mathcal{F}_\lambda(\mathbb{RP}^1)$ *and the operators of degree k given, in an affine coordinate, by the formula*

$$\mathrm{D}_k : \phi(x)(\mathrm{d}x)^{(1-k)/2} \mapsto \frac{\mathrm{d}^k\phi(x)}{\mathrm{d}x^k}(\mathrm{d}x)^{(1+k)/2}. \tag{2.3}$$

Proof The action of SL(2, $\mathbb{R}$) is given, in an affine chart, by the formula

$$x \mapsto \frac{ax+b}{cx+d}. \tag{2.4}$$

Exercise 2.1.3 Prove that the operators D_k are PGL(2, $\mathbb{R}$)-invariant.

The infinitesimal version of formula (2.4) gives the action of the Lie algebra sl(2, $\mathbb{R}$).

Exercise 2.1.4 (a) Prove that the sl(2, $\mathbb{R}$)-action on $\mathbb{RP}^1$ is generated by the three vector fields

$$\frac{\mathrm{d}}{\mathrm{d}x}, \quad x\frac{\mathrm{d}}{\mathrm{d}x}, \quad x^2\frac{\mathrm{d}}{\mathrm{d}x}. \tag{2.5}$$

(b) Prove that the corresponding action on $\mathcal{F}_\lambda(\mathbb{RP}^1)$ is given by the following operators (the Lie derivatives):

$$L^\lambda_{\mathrm{d}/\mathrm{d}x} = \frac{\mathrm{d}}{\mathrm{d}x}, \quad L^\lambda_{x\,\mathrm{d}/\mathrm{d}x} = x\frac{\mathrm{d}}{\mathrm{d}x} + \lambda, \quad L^\lambda_{x^2\mathrm{d}/\mathrm{d}x} = x^2\frac{\mathrm{d}}{\mathrm{d}x} + 2\lambda\, x. \tag{2.6}$$

Consider now a differential operator

$$A = a_k(x)\frac{\mathrm{d}^k}{\mathrm{d}x^k} + \cdots + a_0(x)$$

from $\mathcal{F}_\lambda(\mathbb{RP}^1)$ to $\mathcal{F}_\mu(\mathbb{RP}^1)$ and assume that A is SL(2, $\mathbb{R}$)-invariant. This means that

$$A \circ L^\lambda_X = L^\mu_X \circ A$$

for all $X \in$ sl(2, $\mathbb{R}$).

Take $X = \mathrm{d}/\mathrm{d}x$ to conclude that all the coefficients $a_i(x)$ of A are constants. Now take $X = x\mathrm{d}/\mathrm{d}x$:

$$L^{\mu}_{x\,\mathrm{d}/\mathrm{d}x} \circ \left(\sum a_i \frac{\mathrm{d}^i}{\mathrm{d}x^i}\right) = \left(\sum a_i \frac{\mathrm{d}^i}{\mathrm{d}x^i}\right) \circ L^{\lambda}_{x\,\mathrm{d}/\mathrm{d}x}.$$

Using (2.6), it follows that $a_i(i + \lambda - \mu) = 0$ for all i. Hence all a_i but one vanish, and $\mu = \lambda + k$ where k is the order of A.

Finally, take $X = x^2\mathrm{d}/\mathrm{d}x$. One has

$$L^{\lambda+k}_{x^2\mathrm{d}/\mathrm{d}x} \circ \frac{\mathrm{d}^k}{\mathrm{d}x^k} = \frac{\mathrm{d}^k}{\mathrm{d}x^k} \circ L^{\lambda}_{x^2\mathrm{d}/\mathrm{d}x}.$$

If $k \geq 1$ then, using (2.6) once again, one deduces that $2\lambda = 1 - k$, as claimed; if $k = 0$, then A is proportional to the identity. □

The operator D_1 is just the differential of a function. This is the only operator invariant under the full group Diff($\mathbb{RP}^1$). The operator D_2 is a Sturm–Liouville operator already introduced in Section 1.3. Such an operator determines a projective structure on $\mathbb{RP}^1$. Not surprisingly, the projective structure, corresponding to D_2, is the standard projective structure whose symmetry group is PGL(2, $\mathbb{R}$). The geometric meaning of the operators D_k with $k \geq 3$ will be discussed in the next section.

Comment

The classification problem of invariant differential operators was posed by Veblen in his talk at the International Congress of Mathematicians (ICM) in 1928. Many important results have been obtained since then. The only unary invariant differential operator on tensor fields (and tensor densities) is the de Rham differential; cf. [114, 179].

2.2 Curves in $\mathbb{RP}^n$ and linear differential operators

In Sections 1.3 and 1.4 we discussed the relations between non-degenerate curves and linear differential operators in dimensions 1 and 2. In this section we will extend this construction to the multi-dimensional case.

Constructing differential operators from curves

We associate a linear differential operator

$$A = \frac{\mathrm{d}^{n+1}}{\mathrm{d}x^{n+1}} + a_{n-1}(x)\frac{\mathrm{d}^{n-1}}{\mathrm{d}x^{n-1}} + \cdots + a_1(x)\frac{\mathrm{d}}{\mathrm{d}x} + a_0(x) \qquad (2.7)$$

with a non-degenerate parameterized curve $\gamma(x)$ in $\mathbb{RP}^n$. Consider a lift $\Gamma(x)$ of the curve $\gamma(x)$ to $\mathbb{R}^{n+1}$. Since γ is non-degenerate, the Wronski determinant

$$W(x) = |\Gamma(x), \Gamma'(x), \dots, \Gamma^{(n)}(x)|$$

does not vanish. Therefore the vector $\Gamma^{(n+1)}$ is a linear combination of Γ, Γ', $\dots, \Gamma^{(n)}$, more precisely,

$$\Gamma^{(n+1)}(x) + \sum_{i=0}^{n} a_i(x)\Gamma^{(i)}(x) = 0.$$

This already gives us a differential operator depending, however, on the lift.

Let us find a new, canonical, lift for which the Wronski determinant identically equals 1. Any lift of $\gamma(x)$ is of the form $\alpha(x)\Gamma(x)$ for some non-vanishing function $\alpha(x)$. The condition on this function is

$$|\alpha\Gamma, (\alpha\Gamma)', \dots, (\alpha\Gamma)^{(n)}| = 1,$$

and hence

$$\alpha(x) = W(x)^{-1/(n+1)}. \tag{2.8}$$

For this lift the coefficient $a_n(x)$ in the preceding formula vanishes and the corresponding operator is of the form (2.7). This operator is uniquely defined by the curve $\gamma(x)$.

Exercise 2.2.1 Prove that two curves define the same operator (2.7) if and only if they are projectively equivalent.

Hint. The "if" part follows from the uniqueness of the canonical lift of the projective curve. The "only if" part is more involved and is discussed throughout this section.

Tensor meaning of the operator A and $\mathrm{Diff}(S^1)$-action

Let us discuss how the operator A depends on the parameterization of the curve $\gamma(x)$. The group $\mathrm{Diff}(S^1)$ acts on parameterized curves by reparameterization. To a parameterized curve we assigned a differential operator (2.7). Thus one has an action of $\mathrm{Diff}(S^1)$ on the space of such operators. We call it the *geometric action*.

Let us define another, *algebraic action* of $\mathrm{Diff}(S^1)$ on the space of operators (2.7)

$$A \mapsto \mathrm{T}_f^{(n+2)/2} \circ A \circ \mathrm{T}_{f^{-1}}^{-n/2}, \qquad f \in \mathrm{Diff}(S^1), \tag{2.9}$$

which is, of course, a particular case of the action of Diff(S^1) on $\mathcal{D}_{\lambda,\mu}(S^1)$ as in (2.1). In other words, $A \in \mathcal{D}^{n+1}_{-n/2,(n+2)/2}(S^1)$.

Exercise 2.2.2 The action of Diff(S^1) on $\mathcal{D}_{\lambda,\mu}(S^1)$ preserves the specific form of the operators (2.7), namely, the highest-order coefficient equals 1 and the next highest equals zero, if and only if

$$\lambda = -\frac{n}{2} \qquad \text{and} \qquad \mu = \frac{n+2}{2}.$$

Theorem 2.2.3 *The two* Diff(S^1)*-actions on the space of differential operators (2.7) coincide.*

Proof Let us start with the geometric action. Consider a new parameter $y = f(x)$ on γ. Then

$$\Gamma_x = \Gamma_y f', \quad \Gamma_{xx} = \Gamma_{yy}(f')^2 + \Gamma_y f'',$$

etc., where Γ is a lifted curve and f' denotes $\mathrm{d}f/\mathrm{d}x$. It follows that

$$|\Gamma, \Gamma_x, \dots, \Gamma_{x\dots x}| = |\Gamma, \Gamma_y, \dots, \Gamma_{y\dots y}|(f')^{n(n+1)/2},$$

and, therefore, the Wronski determinant $W(x)$ is a tensor density of degree $n(n+1)/2$, that is, an element of $\mathcal{F}_{n(n+1)/2}$. Hence the coordinates of the canonical lift $\alpha\Gamma$ given by (2.8) are tensor densities of degree $-n/2$ (we already encountered a particular case $n = 2$ in Exercise 1.6.5). Being the coordinates of the canonical lift $\alpha\Gamma$, the solutions of the equation

$$A\phi = 0 \tag{2.10}$$

are $(-n/2)$-densities.

From the very definition of the algebraic action (2.9) it follows that the kernel of the operator A consists of $(-n/2)$-densities. It remains to note that the kernel uniquely defines the corresponding operator. □

The brevity of the proof might be misleading. An adventurous reader may try to prove Theorem 2.2.3 by a direct computation. Even for the Sturm–Liouville ($n = 2$) case this is quite a challenge (see, e.g., [37]).

Example 2.2.4 The SL(2, $\mathbb{R}$)-invariant linear differential operator (2.3) fits into the present framework. This operator corresponds to a remarkable parameterized projective curve in $\mathbb{RP}^{k-1}$, called the *normal curve*, uniquely characterized by the following property. The parameter x belongs to S^1 and corresponds to the canonical projective structure on S^1. If one changes the parameter by a linear-fractional transformation $x \mapsto (ax + b)/(cx + d)$, the resulting curve is projectively equivalent to the original one. In appropriate affine coordinates, this curve is given by

$$\gamma = (1 : x : x^2 : \cdots : x^{k-1}). \tag{2.11}$$

Dual operators and dual curves

Given a linear differential operator $A : \mathcal{F}_\lambda \to \mathcal{F}_\mu$ on S^1, its dual operator $A^* : \mathcal{F}_{1-\mu} \to \mathcal{F}_{1-\lambda}$ is defined by the equality

$$\int_{S^1} A(\phi)\psi = \int_{S^1} \phi A^*(\psi)$$

for any $\phi \in \mathcal{F}_\lambda$ and $\psi \in \mathcal{F}_{1-\mu}$. The operation $A \mapsto A^*$ is Diff(S^1)-invariant. An explicit expression for the dual operator was already given (1.18).

If $\lambda + \mu = 1$ then the operator A^* has the same domain and the same range as A. In this case, there is a decomposition

$$A = \left(\frac{A + A^*}{2}\right) + \left(\frac{A - A^*}{2}\right)$$

into the symmetric and skew-symmetric parts.

Now let $A \in \mathcal{D}^{n+1}_{-n/2,(n+2)/2}(S^1)$ be the differential operator (2.7) constructed from a projective curve $\gamma(x)$. The modules $\mathcal{F}_{-n/2}$ and $\mathcal{F}_{(n+2)/2}$ are dual to each other. Therefore A can be decomposed into the symmetric and skew-symmetric parts, and this decomposition is independent of the choice of the parameter on the curve. This fact was substantially used in the proof of Theorem 1.4.3.

Consider a projective curve $\gamma(x) \subset \mathbb{RP}^n$, its canonical lift $\Gamma(x) \subset \mathbb{R}^{n+1}$ and the respective differential operator A. The coordinates of the curve Γ satisfy equation (2.10). These coordinates are linear functions on $\mathbb{R}^{n+1}$. Thus the curve Γ lies in the space, dual to ker A, and so ker A is identified with $\mathbb{R}^{n+1*}$.

Now let us define a smooth parameterized curve $\tilde{\Gamma}(x)$ in $\mathbb{R}^{n+1*}$. Given a value of the parameter x, consider the solution ϕ_x of equation (2.10) satisfying the following n initial conditions:

$$\phi_x(x) = \phi_x'(x) = \ldots = \phi_x^{(n-1)}(x) = 0; \tag{2.12}$$

such a solution is unique up to a multiplicative constant. The solution ϕ_x is a vector in $\mathbb{R}^{n+1*}$, and we set $\tilde{\Gamma}(x) = \phi_x$. Define the projective curve $\tilde{\gamma}(x) \subset \mathbb{RP}^{n*}$ as the projection of $\tilde{\Gamma}(x)$.

Exercise 2.2.5 Prove that the curve $\tilde{\gamma}$ coincides with the projectively dual curve γ^*.

Dual curves correspond to dual differential operators

We have two notions of duality, one for projective curves and one for differential operators. The next classical result shows that the two agree.

Theorem 2.2.6 *Let A be the differential operator corresponding to a non-degenerate projective curve $\gamma(x) \subset \mathbb{RP}^n$. Then the differential operator corresponding to the projectively dual curve $\gamma^*(x)$ is $(-1)^{n+1}A^*$.*

Proof Let $U = \text{Ker } A,\ V = \text{Ker } A^*$. We will construct a non-degenerate pairing between these spaces.

Let ϕ and ψ be $(-n/2)$-densities. The expression

$$A(\phi)\psi - \phi A^*(\psi) \tag{2.13}$$

is a differential 1-form on S^1. The integral of (2.13) vanishes, and hence there exists a function $B(\phi, \psi)(x)$ such that

$$A(\phi)\psi - \phi A^*(\psi) = B'(\phi, \psi)\mathrm{d}x. \tag{2.14}$$

If A is given by (2.7) then

$$B(\phi, \psi) = \phi^{(n)}\psi - \phi^{(n-1)}\psi' + \cdots + (-1)^n\phi\psi^{(n)} + b(\phi, \psi),$$

where b is a bidifferential operator of degree $\le n - 1$.

If $\phi \in U$ and $\psi \in V$ then the left hand side of (2.14) vanishes, and therefore $B(\phi, \psi)$ is a constant. It follows that B determines a bilinear pairing of spaces U and V.

The pairing B is non-degenerate. Indeed, fix a parameter value $x = x_0$, and choose a special basis $\phi_0, \ldots, \phi_n \in U$ such that $\phi_i^{(j)}(x_0) = 0$ for all $i \ne j;\ i, j = 0, \ldots, n$, and $\phi_i^{(i)}(x_0) = 1$ for all i. Let $\psi_i \in V$ be the basis in V defined similarly. In these bases, the matrix of $B(\phi, \psi)(x_0)$ is triangular with the diagonal elements equal to ± 1.

The pairing B allows us to identify U^* with V. Consider the curve $\tilde{\Gamma}(x)$ associated with the operator A; this curve belongs to U and consists of solutions (2.12). Let $\widehat{\Gamma}(x) \subset V$ be a similar curve corresponding to A^*. We want to show that these two curves are dual with respect to the pairing B, that is,

$$B(\tilde{\Gamma}^{(i)}(x_0), \widehat{\Gamma}(x_0)) = 0, \quad i = 0, \ldots, n-1 \tag{2.15}$$

for all parameter values x_0.

Indeed, the vector $\tilde{\Gamma}^{(i)}(x_0)$ belongs to the space of solutions U and this solution vanishes at x_0. The function $\widehat{\Gamma}(x_0) \in V$ vanishes at x_0 with the first $n - 1$ derivatives, and (2.15) follows from the above expression for the operator B.

Therefore $\tilde{\gamma}^* = \widehat{\gamma}$, that is, the curves corresponding to A and A^* are projectively dual. □

Exercise 2.2.7 Prove the following explicit formula:

$$B(\phi, \psi) = \sum_{r+s+t \leq n} (-1)^{r+t+1} \binom{r+t}{r} a_{r+s+t+1}^{(r)} \phi^{(s)} \psi^{(t)}.$$

Remark 2.2.8 If A is a symmetric operator, $A^* = A$, then B is a non-degenerate skew-symmetric bilinear form, that is, a symplectic structure, on the space Ker A – cf. [166].

Monodromy

If γ is a closed curve, then the operator A has periodic coefficients. The converse is not at all true. Let A be an operator with periodic coefficients, in other words, a differential operator on S^1. The solutions of the equation $A\phi = 0$ are not necessarily periodic; they are defined on $\mathbb{R}$, viewed as the universal covering of $S^1 = \mathbb{R}/2\pi\mathbb{Z}$. One obtains a linear map on the space of solutions:

$$\mathrm{T} : \phi(x) \mapsto \phi(x + 2\pi)$$

called the monodromy. Monodromy was already mentioned in Sections 1.3 and 1.6.

Consider in more detail the case of operators (2.7). The Wronski determinant of any $(n+1)$-tuple of solutions is constant. This defines a volume form on the space of solutions. Since T preserves the Wronski determinant, the monodromy belongs to $\mathrm{SL}(n+1, \mathbb{R})$. Note, however, that this element of $\mathrm{SL}(n+1, \mathbb{R})$ is defined up to a conjugation, for there is no natural basis in ker A and there is no way to identify ker A with $\mathbb{R}^{n+1}$; only a conjugacy class of T has an invariant meaning.

Consider a projective curve $\gamma(x)$ associated with a differential operator A on S^1. Let $\Gamma(x) \subset \mathbb{R}^{n+1}$ be the canonical lift of $\gamma(x)$. Both curves are not necessarily closed, but satisfy the monodromy condition

$$\gamma(x + 2\pi) = \mathrm{T}(\gamma(x)), \qquad \Gamma(x + 2\pi) = \mathrm{T}(\Gamma(x)),$$

where T is a representative of a conjugacy class in $\mathrm{SL}(n+1, \mathbb{R})$.

As a consequence of Theorem 2.2.3, asserting the coincidence of the algebraic $\mathrm{Diff}(S^1)$-action (2.9) on the space of differential operators with the geometric action by reparameterization, we have the following statement.

Corollary 2.2.9 *The conjugacy class in* $\mathrm{SL}(n+1,\mathbb{R})$ *of the monodromy of a differential operator (2.7) is invariant with respect to the* $\mathrm{Diff}_+(S^1)$*-action, where* $\mathrm{Diff}_+(S^1)$ *is the connected component of* $\mathrm{Diff}(S^1)$.

Comment

Representation of parameterized non-degenerate curves in $\mathbb{RP}^n$ (modulo equivalence) by linear differential operators was a basic idea of projective differential geometry of the second half of nineteenth century. We refer to Wilczynski's book [231] for a first systematic account of this approach. Our proof of Theorem 2.2.6 follows that of [231] and [13]; a different proof can be found in [106].

2.3 Homotopy classes of non-degenerate curves

Differential operators on $\mathbb{RP}^1$ of the special form (2.7) correspond to non-degenerate curves in $\mathbb{RP}^n$. In this section we give a topological classification of such curves. We study homotopy equivalence classes of non-degenerate immersed curves with respect to the homotopy, preserving the non-degeneracy. This allows us to distinguish interesting classes of curves, such as that of convex curves.

Curves in S^2: a theorem of J. Little

Let us start with the simplest case, the classification problem for non-degenerate curves on the 2-sphere.

Theorem 2.3.1 *There are three homotopy classes of non-degenerate immersed non-oriented closed curves on* S^2 *represented by the curves in Figure 2.1.*

Proof Recall the classical Whitney theorem on the regular homotopy classification of closed plane immersed curves. To such a curve one assigns the winding number: a non-negative integer equal to the total number of turns of the tangent line (see Figure 2.2). The curves are regularly homotopic if and only if

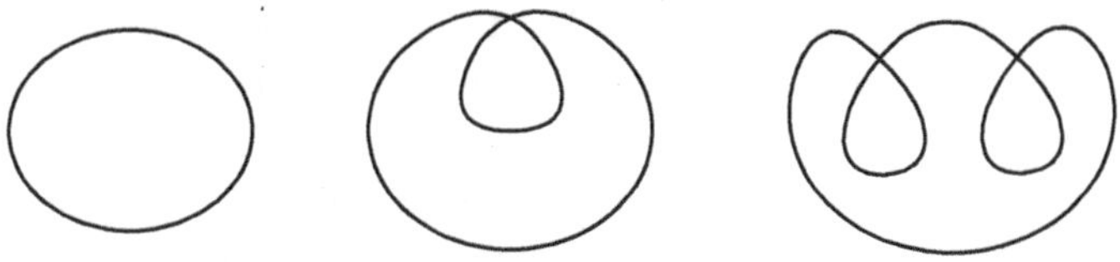

Fig. 2.1. Non-degenerate curves on S^2

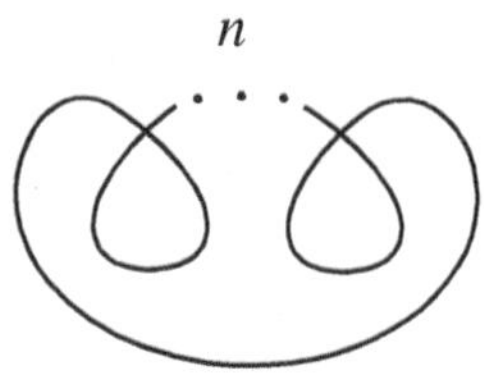

Fig. 2.2. Winding number n

their winding numbers are equal. The spherical version of the Whitney theorem is simpler: there are only two regular homotopy classes of closed immersed curves on S^2, represented by the first and the second curves in Figure 2.1. The complete invariant is the parity of the number of double points.

The Whitney theorem extends to non-degenerate plane curves and the proof dramatically simplifies.

Lemma 2.3.2 *The winding number is the complete invariant of non-degenerate plane curves with respect to non-degenerate homotopy.*

Proof A non-degenerate plane curve can be parameterized by the angle made by the tangent line with a fixed direction. In such a parameterization, a linear homotopy connects two curves with the same winding number. □

We are ready to proceed to the proof of Theorem 2.3.1.

Part 1. Let us prove that the three curves in Figure 2.1 are not homotopic as non-degenerate curves. The second curve is not even regularly homotopic to the other two. We need to prove that the first and third curves are not homotopic.

The first curve is *convex*: it intersects any great circle at at most two points. We understand intersections in the algebraic sense, that is, with multiplicities. For example, the curve $y = x^2$ has double intersection with the x-axis.

Lemma 2.3.3 *A convex curve remains convex under homotopies of non-degenerate curves.*

Proof Arguing by contradiction, assume that there is a homotopy destroying convexity. Convexity is an open condition. Consider the first moment when the curve fails to be convex. At this moment, there exists a great circle intersecting the curve with total multiplicity four. The following three cases are possible: (a) four distinct transverse intersections, (b) two transverse and one tangency, (c) two tangencies; see Figure 2.3. Note that a non-degenerate curve cannot have intersection multiplicity >2 with a great circle at a point.

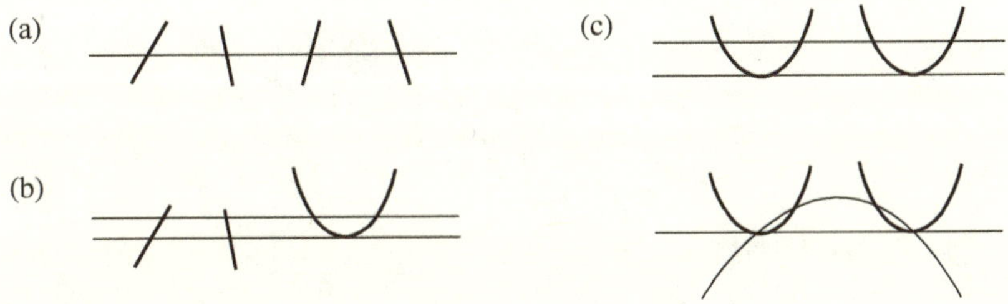

Fig. 2.3. Total multiplicity 4

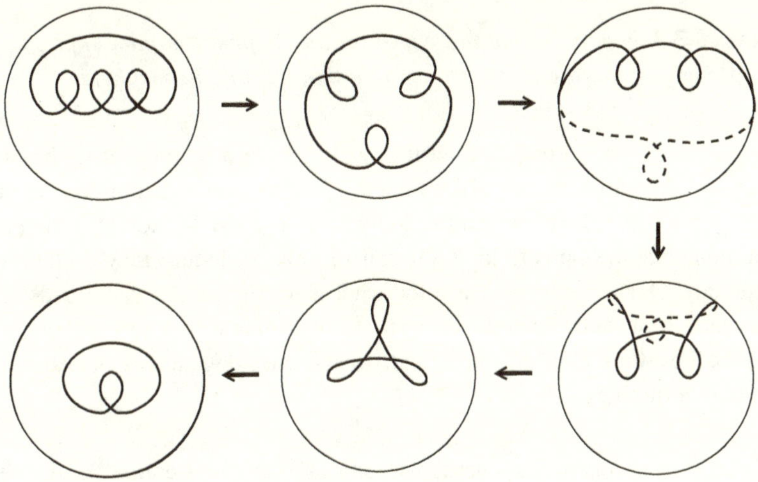

Fig. 2.4. Homotopy 4 ⟶ 2

Case (a) is impossible: since transverse intersection is an open condition, this cannot be the first moment when the curve fails to be convex. In cases (b) and (c) one can perturb the great circle so that the intersections become transverse and we are back to case (a). In case (b) this is obvious, as well as in case (c) if the two points of tangency are not antipodal; see Figure 2.3. For antipodal points, one rotates the great circle about the axis connecting the tangency points. □

Part 2. Let us now prove that a non-degenerate curve on S^2 is non-degenerate homotopic to a curve in Figure 2.1. Unlike the planar case, non-degenerate curves with winding number n and $n+2$, where $n \geq 2$, are non-degenerate homotopic; see Figure 2.4. The apparent inflection points are not really there; see [137] for a motion picture featuring front-and-back view. The authors recommend the reader to repeat their experience and to draw the picture on a well-inflated ball.

Fig. 2.5. Spiral curve

Therefore, any curve in Figure 2.2 is, indeed, homotopic to a curve in Figure 2.1 in the class of non-degenerate curves.

Lemma 2.3.4 *A non-degenerate curve on the 2-sphere is homotopic, in the class of non-degenerate curves, to a curve that lies in a hemisphere.*

Proof If the curve is convex, then it already lies in a hemisphere. The proof of this fact is similar to that of Lemma 2.3.3. If the curve is not convex, then pick a point p off the curve and radially homotop (by homothety) the curve to the hemisphere opposite to p. Of course, this procedure may violate non-degeneracy. Our remedy is to use the trick from Figure 2.4 backwards. This allows us to replace the original curve by a homotopic C^0-close spiral curve, as in Figure 2.5. If the kinks are sufficiently dense, non-degeneracy persists under the radial homothety. □

Given a non-degenerate curve, we may assume, by Lemma 2.3.4, that it lies in a hemisphere. Identifying the hemisphere with the plane by the central projection, the non-degeneracy is preserved. Thus, we have a non-degenerate curve in the plane and, by Lemma 2.3.2, we may assume that it is one of the curves in Figure 2.2. And finally, the trick from Figure 2.4 reduces the winding number to $n = 1, 2, 3$. This concludes the proof of the theorem. □

Remark 2.3.5 Let us emphasize that Theorem 2.3.1 concerns non-oriented curves. Taking orientation into consideration, the number of homotopy classes is equal to 6.

Multi-dimensional case and quasi-periodic curves

We are essentially concerned with curves in $\mathbb{RP}^n$. Little's theorem provides a classification of non-degenerate closed curves in $\mathbb{RP}^2$ and the number of homotopy classes of such curves is again 3.

Exercise 2.3.6 Show that a non-degenerate closed curve in $\mathbb{RP}^2$ is contractible.

Therefore, a non-degenerate closed curve in $\mathbb{RP}^2$ lifts to S^2 as a closed curve, and one applies the Little theorem.

The most general result in the multi-dimensional case is contained in the following theorem.

Theorem 2.3.7 *(i) There are three homotopy classes of non-degenerate immersed non-oriented closed curves on S^{2n} and two such classes on S^{2n+1}, where $n \geq 1$.*
(ii) There are three homotopy classes of non-degenerate immersed non-oriented closed curves in $\mathbb{R}^{2n-1}$ and two such classes in $\mathbb{R}^{2n}$, where $n \geq 2$.
(iii) There are three homotopy classes of non-degenerate immersed non-oriented closed curves in $\mathbb{RP}^{2n}$ and five such classes in $\mathbb{RP}^{2n+1}$, where $n \geq 1$.

We do not prove this theorem; let us, however, explain the origin of the somewhat surprising answer 5 in case (iii). Unlike the even-dimensional situation, a non-degenerate curve in $\mathbb{RP}^{2n+1}$ can be non-contractible. The canonical lift of such a curve to $\mathbb{R}^{2n+2}$ has monodromy $-\mathbf{1}$. The number of homotopy classes splits as follows: $5 = 2 + (2 + 1)$. The first two classes consist of contractible non-degenerate curves with the unique invariant interpreted as an element of $\pi_1(\mathrm{SO}(2n + 2)) = \mathbb{Z}_2$. The remaining three classes consist of non-contractible curves with the same invariant in $\mathbb{Z}_2$ and a distinguished class of *convex* (or disconjugate) curves in $\mathbb{RP}^n$, analogous to convex ones in Little's theorem.

The case of non-degenerate curves with monodromy was studied only in the two-dimensional case. The answer here is either 2 or 3, depending on the monodromy.

Comment

Theorem 2.3.1 was proved in [137]. This work was not related to differential operators and extended a previous work by W. Pohl. Our proof is new and simpler then the original one. The idea of curling the curve in our proof of Lemma 2.3.4 resembles Thurston's proof of the Whitney theorem in the movie "Outside in" [160]. However the core component of the proof, the homotopy in Figure 2.4, is due to Little.

Theorem 2.3.7 is due to M. Shapiro; see [189]. Part (ii) is a generalization of [138]. For the case of non-trivial monodromy see [107].

2.4 Two differential invariants of curves: projective curvature and cubic form

In Section 1.4 we associated two differential invariants with a non-degenerate curve γ in $\mathbb{RP}^2$. The first is called the projective curvature; it is defined as a projective structure on γ. The second one is a cubic form on γ. In this section we will generalize these two invariants to curves in $\mathbb{RP}^n$.

Projective curvature

Given a non-degenerate curve $\gamma \subset \mathbb{RP}^n$, we will define a projective structure on γ, invariant with respect to projective transformations of $\mathbb{RP}^n$.

Choose an arbitrary parameter x on γ and associate a linear differential operator (2.7) with the parameterized curve $\gamma(x)$. The main ingredient of the construction is the following observation: there is a natural projection from the space of operators (2.7) to the space of Sturm–Liouville operators.

Theorem 2.4.1 *The map $A \mapsto L$ associating the Sturm–Liouville operator*

$$L = c\,\frac{\mathrm{d}^2}{\mathrm{d}x^2} + a_{n-1}(x), \qquad \textit{where} \quad c = \binom{n+2}{3}, \tag{2.16}$$

with a differential operator (2.7) is $\mathrm{Diff}(S^1)$*-invariant.*

Proof The $\mathrm{Diff}(S^1)$-action on the space of Sturm–Liouville operators is given by the formula (2.12). We need to check that the map $A \mapsto L$ commutes with the $\mathrm{Diff}(S^1)$-action.

Exercise 2.4.2 Check by a direct computation that the result of the $\mathrm{Diff}(S^1)$-action (2.9) is a linear differential operator with the $(n-1)$-order coefficient

$$\left(\frac{\mathrm{d}f^{-1}}{\mathrm{d}t}\right)^2 a_{n-1}(f^{-1}) + \frac{c}{2}\, S(f^{-1}),$$

where c is as in (2.16).

We have already encountered a similar formula: this is the $\mathrm{Diff}(S^1)$-action on the coefficient of a Sturm–Liouville operator $L = c(d/dx)^2 + a_{n-1}$; cf. formula (2.12) in which, however, $c = 1$. □

Projective structures on S^1 are identified with Sturm–Liouville operators (see Section 1.3), so that the operator (2.16) defines a projective structure on $\gamma(x)$.

Theorem 2.4.1 implies that this projective structure does not depend on the parameterization.

Remark 2.4.3 (a) A particular case, $n = 2$, of the above correspondence was considered in detail in Section 1.4. Let us emphasize once again that this notion cannot be understood as a function or a tensor field.

(b) Theorem 2.4.1 implies that there exists a (local) parameter x such that the coefficient $a_{n-1}(x)$ of the operator (2.7) is identically zero. Indeed, choose a local coordinate of the projective structure characterized by the property that the potential of the corresponding Sturm–Liouville operator vanishes in this coordinate. This special form of the operator (2.7) is called the Forsyth–Laguerre canonical form. Note that such a parameter x is defined up to a linear-fractional transformation.

Cubic form

Another important invariant of a non-degenerate projective curve γ is a cubic form on γ.

As above, choose a parameter x on γ and consider the corresponding differential operator A given by (2.7). Consider a (skew-) symmetric operator

$$A_0 = \tfrac{1}{2}\left(A + (-1)^n A^*\right) \tag{2.17}$$

from $\mathcal{F}_{-n/2}$ to $\mathcal{F}_{1+n/2}$.

The principal symbol $\sigma(A_0)$ is a cubic form on γ. Indeed, the principal symbol of a differential operator from $\mathcal{D}^k_{\lambda,\mu}(S^1)$ is a tensor density of degree $\mu - \lambda - k$; see Exercise 2.1.1. Since $A_0 \in \mathcal{D}^{n-2}_{-n/2,1+n/2}(S^1)$, its highest-order coefficient is a tensor density on S^1 of degree 3; the parameter x pushes forward this 3-density to γ.

Exercise 2.4.4 Check that the operator (2.17) is of order $n - 2$ and its highest-order coefficient is proportional to

$$\tau(x) = a_{n-2}(x) - \frac{n-1}{2}\, a'_{n-1}(x). \tag{2.18}$$

The formula $\tau(x)(\mathrm{d}x)^3$, where $\tau(x)$ is as in (2.18), can be chosen as an alternative definition of the cubic form on γ. One can check directly that it, indeed, transforms with respect to the $\mathrm{Diff}(S^1)$-action as a cubic form.

Remark 2.4.5 Note that one can construct another tensor invariant from the Sturm–Liouville operator (2.16) and the cubic form (2.18):

$$\theta = \left(6\tau\tau'' - 7(\tau')^2 - \frac{18}{c}\,a_{n-1}\tau^2\right)(\mathrm{d}x)^8,$$

where, as before, $c = \binom{n+2}{3}$.

Comment

Projective curvature and the cubic form are the simplest invariants of a curve in projective space (see Section 1.4 for historical comments). For the plane curves they form a complete set of invariants.

2.5 Projectively equivariant symbol calculus

In this section we study the space $\mathcal{D}_{\lambda,\mu}(S^1)$ of linear differential operators on tensor densities on S^1 as a PGL(2, $\mathbb{R}$)-module. We define a canonical PGL(2, $\mathbb{R}$)-isomorphism between this space and the space of tensor densities on S^1. We apply this isomorphism to construct projective differential invariants of nondegenerate curves in $\mathbb{RP}^n$.

Space of symbols

There is a natural filtration

$$\mathcal{D}^0_{\lambda,\mu}(S^1) \subset \mathcal{D}^1_{\lambda,\mu}(S^1) \subset \cdots \subset \mathcal{D}^k_{\lambda,\mu}(S^1) \subset \cdots.$$

The corresponding graded Diff(S^1)-module $\mathcal{S}_{\lambda,\mu}(S^1) = \mathrm{gr}\,(\mathcal{D}_{\lambda,\mu}(S^1))$ is called the *module of symbols of differential operators.*

The quotient-module $\mathcal{D}^k_{\lambda,\mu}(S^1)/\mathcal{D}^{k-1}_{\lambda,\mu}(S^1)$ is isomorphic to the module of tensor densities $\mathcal{F}_{\mu-\lambda-k}(S^1)$; the isomorphism is provided by the principal symbol. Therefore, as a Diff(S^1)-module, the space of symbols depends only on the difference

$$\delta = \mu - \lambda,$$

so that $\mathcal{S}_{\lambda,\mu}(S^1)$ can be written as $\mathcal{S}_\delta(S^1)$, and finally we have

$$\mathcal{S}_\delta(S^1) = \bigoplus_{k=0}^{\infty} \mathcal{F}_{\delta-k}(S^1)$$

as Diff(S^1)-modules.

PGL(2, $\mathbb{R}$)-equivariant "total" symbol map

The principal symbol map is a Diff(S^1)-equivariant projection $\sigma : \mathcal{D}^k_{\lambda,\mu}(S^1) \to \mathcal{F}_{\delta-k}(S^1)$. We would like to identify the full space of differential operators with the total space of symbols $\mathcal{S}_\delta(S^1)$. That is, we are looking for a linear bijection

$$\sigma_{\lambda,\mu} : \mathcal{D}_{\lambda,\mu}(S^1) \xrightarrow{\simeq} \mathcal{S}_\delta(S^1) \tag{2.19}$$

such that the highest-order term of $\sigma_{\lambda,\mu}(A)$ coincides with the principal symbol $\sigma(A)$ for all $A \in \mathcal{D}_{\lambda,\mu}(S^1)$. Such an isomorphism is usually called a *symbol map*.

Ideally, this map should commute with the Diff(S^1)-action; however, this is not possible for cohomological reasons (see [133]). It turns out that there is a unique map (2.19), equivariant with respect to the subgroup of projective (linear-fractional) transformations PGL(2, $\mathbb{R}$) $\subset$ Diff(S^1).

Theorem 2.5.1 *There exists a unique* PGL(2, $\mathbb{R}$)-*invariant symbol map (2.19), provided*

$$\delta \notin \left\{1, \tfrac{3}{2}, 2, \tfrac{3}{2}, \ldots\right\}. \tag{2.20}$$

It sends a differential operator $A = \sum_j a_j(x)(\mathrm{d}/\mathrm{d}x)^j$ *to the tensor density*

$$\sigma_{\lambda,\mu}(A) = \sum_{j\geq 0}\sum_{\ell=0}^{j} C^j_\ell \, a_j^{(\ell)}(x)\,(\mathrm{d}x)^{\delta-j+\ell}, \tag{2.21}$$

where $a_j^{(\ell)} = \mathrm{d}^\ell a_j(x)/\mathrm{d}x^\ell$ *and* C^j_ℓ *are the following constants*

$$C^j_\ell = (-1)^\ell \frac{\binom{j}{\ell}\binom{j+2\lambda-1}{\ell}}{\binom{2j-\ell-2\delta+1}{\ell}}. \tag{2.22}$$

Proof The proof consists of two steps. First, one proves that any PGL(2, $\mathbb{R}$)-invariant linear map (2.19) is given by a differential operator, that is, is of the form

$$\sigma_{\lambda,\mu}(A) = \sum_{j,\ell,m} C^j_{\ell,m}(x)\, a_j^{(\ell)}(x)(\mathrm{d}x)^m,$$

where $C^j_{\ell,m}(x)$ are smooth functions on S^1. Second, one shows that such an operator, commuting with the PGL(2, $\mathbb{R}$)-action and preserving the principal symbol, is defined by (2.21) and (2.22).

We omit here the first step (see [134]). The second part of the proof is similar to those of Theorems 2.1.2 and 3.1.1. Consider the action of the Lie algebra sl(2, $\mathbb{R}$) generated by the vector fields $\mathrm{d}/\mathrm{d}x$, $x\mathrm{d}/\mathrm{d}x$ and $x^2\mathrm{d}/\mathrm{d}x$.

(a) Equivariance with respect to the vector field $\mathrm{d}/\mathrm{d}x$ implies that all the coefficients $C^j_{\ell,m}(x)$ in $\sigma_{\lambda,\mu}$ are constants.
(b) Equivariance with respect to the vector field $x\,\mathrm{d}/\mathrm{d}x$ implies the homogeneity condition $j - \ell - m = 0$, so that $\sigma_{\lambda,\mu}$ is of the form (2.21).
(c) Consider the third vector field $x^2\mathrm{d}/\mathrm{d}x$.

Exercise 2.5.2 Check by a straightforward computation that equivariance with respect to the vector field $x^2\mathrm{d}/\mathrm{d}x$ leads to the recurrence relation

$$C^{j+1}_{\ell+1} = \frac{(j+1)(j+2\lambda)}{(\ell+1)(\ell+2(\delta-j-1))}\, C^j_\ell.$$

This relation, together with the normalization condition $C^j_0 = 1$, equivalent to the fact that the principal symbol is preserved, readily gives the solution (2.22). The restriction (2.20) guarantees that the denominator in the recurrence formula does not vanish. □

Hypergeometric function

The symbol map defined by (2.21) and (2.22) can by written in a more elegant way. Let us use the following standard notation: a differential operator $A = \sum_j a_j(x)(\mathrm{d}/\mathrm{d}x)^j$ is denoted by $\sum_j a_j(x)\xi^j$, where ξ is a formal variable. Introduce two differential operators:

$$\mathcal{E} = \xi\frac{\partial}{\partial\xi}, \qquad \mathrm{D} = \frac{\partial}{\partial x}\frac{\partial}{\partial\xi},$$

the Euler field and the "divergence". It turns out that the total symbol map $\sigma_{\lambda,\mu}$ can be expressed in terms of $\mathcal{E}$ and D.

A confluent hypergeometric function of one variable z and two parameters a, b is defined by the series

$$F\left(\begin{matrix} a \\ b \end{matrix}\middle| z\right) = \sum_{m=0}^{\infty} \frac{(a)_m}{(b)_m}\frac{z^m}{m!} \qquad (2.23)$$

where $(a)_m = a(a+1)\cdots(a+m-1)$.

Exercise 2.5.3 Check that the total symbol map, given by (2.21) and (2.22), can be written as a hypergeometric function (2.23) with

$$a = \mathcal{E} + 2\lambda, \quad b = 2(\mathcal{E} - \delta + 1) \quad \text{and} \quad z = -\mathrm{D}.$$

Projectively equivariant quantization map

It is also useful to consider the *quantization map*, which is the inverse of the symbol map:

$$\mathcal{Q}_{\lambda,\mu} = {\sigma_{\lambda,\mu}}^{-1}. \tag{2.24}$$

Let us give here the explicit formula for $\mathcal{Q}_{\lambda,\mu}$.

Theorem 2.5.4 *The quantization map $\mathcal{Q}_{\lambda,\mu}$ is given by the hypergeometric function*

$$F\left(\begin{matrix} a_1 & a_2 \\ b_1 & b_2 \end{matrix}\middle| z\right) = \sum_{m=0}^{\infty} \frac{(a_1)_m\,(a_2)_m}{(b_1)_m\,(b_2)_m}\,\frac{z^m}{m!}$$

where

$$\begin{aligned} a_1 &= \mathcal{E} + 2\lambda, & a_2 &= 2\mathcal{E} - 2(\delta - 1) - 1, \\ b_1 &= \mathcal{E} - \delta + \tfrac{1}{2}, & b_2 &= \mathcal{E} - \delta + 1 \end{aligned}$$

and

$$z = \frac{\mathrm{D}}{4}.$$

The proof is similar to that of Theorem 2.5.1 and Exercise 2.5.3.

Higher differential invariants of non-degenerate curves

Restricting $\sigma_{\lambda,\mu}$ to the subspace $\mathcal{D}^k_{\lambda,\mu}(S^1)$, there is a finite set of exceptional values: $\delta \in \{1, 3/2, \ldots, k+1\}$. Indeed, the denominator of the recurrence formula vanishes for

$$\delta = j - \frac{\ell}{2} + 1 \tag{2.25}$$

where $j \le k$ and $\ell \le j$. In particular, for the module $\mathcal{D}^{n+1}_{-n/2,1+n/2}(S^1)$, the value of $\delta = n+1$ is exceptional. Equation (2.25) has two solutions: ($j = n+1$, $\ell = 2$) and ($j = n$, $\ell = 0$). For these values formula (2.22) is meaningless and, indeed, the symbol map $\sigma_{\lambda,\mu}$ does not exist.

We are, however, concerned with differential operators related to projective curves that have the special form (2.7). Theorem 2.5.1 still applies to the subspace of such differential operators. Since $a_{n+1} = 1$ and $a_n = 0$, the coefficients C^{n+1}_ℓ and C^n_ℓ are irrelevant, except C^{n+1}_0, which equals 1. The coefficients C^j_ℓ with $j \le n-1$ are given by (2.22).

As a consequence of the previous constructions, we obtain higher differential invariants of non-degenerate curves in $\mathbb{RP}^n$. First of all, such a curve carries a canonical projective structure; see Section 2.4. The choice of a projective structure allows us to reduce the infinite-dimensional group $\mathrm{Diff}(S^1)$ to $\mathrm{PGL}(2,\mathbb{R})$. Choosing an adopted local parameter x, one obtains a differential operator $A \in \mathcal{D}^{n+1}_{-n/2,1+n/2}(S^1)$ of the special form (2.7). Now each term of the total symbol (2.21)–(2.22) is a projective differential invariant of the curve. One of these invariants, the cubic form from Section 2.4, is actually invariant under the full group $\mathrm{Diff}(S^1)$.

The constructed set of projective invariants is complete. Indeed, the differential operator A characterizes the curve up to projective equivalence, while the total symbol completely determines the operator.

Comment

The theory of projective differential invariants of non-degenerate curves is a classical subject. It was thoroughly studied in the nineteenth century and summarized as early as in 1906 by Wilczynski [231]. A projectively equivariant symbol map was defined only recently in [44] and [134]; its hypergeometric interpretation was found in [57].

3

The algebra of the projective line and cohomology of Diff(S^1)

Geometry and algebra can be hardly separated: "*L'algèbre n'est qu'une géométrie écrite; la géométrie n'est qu'une algèbre figurée*".[1] Geometric objects usually form algebras, such as Lie algebras of vector fields, associative algebras of differential operators, etc.

In this chapter we consider the associative algebra of differential operators on the projective line. Projective geometry allows us to describe this complicated and interesting object in terms of tensor densities. The group $\mathrm{Diff}_+(S^1)$ and its cohomology play a prominent role, unifying different aspects of our study.

The group $\mathrm{Diff}_+(S^1)$ of orientation-preserving diffeomorphisms of the circle is one of the most popular infinite-dimensional Lie groups connected to numerous topics in contemporary mathematics. The corresponding Lie algebra, $\mathrm{Vect}(S^1)$, also became one of the main characters in various areas of mathematical physics and differential geometry. Part of the interest in the cohomology of $\mathrm{Vect}(S^1)$ and $\mathrm{Diff}(S^1)$ is due to the existence of their non-trivial central extensions, the Virasoro algebra and the Virasoro group.

We consider the first and the second cohomology spaces of $\mathrm{Diff}(S^1)$ and $\mathrm{Vect}(S^1)$ with some non-trivial coefficients and investigate their relations to projective differential geometry. Cohomology of $\mathrm{Diff}(S^1)$ and $\mathrm{Vect}(S^1)$ has been studied by many authors in many different settings; see [72] and [90] for comprehensive references. Why should we consider this cohomology here?

The group $\mathrm{Diff}(S^1)$, the Lie algebra $\mathrm{Vect}(S^1)$ and the Virasoro algebra appear consistently throughout this book. Their relation to the Schwarzian derivative was discussed in Sections 1.3, 1.5 and 1.6. The Schwarzian derivative is a projectively invariant 1-cocycle on $\mathrm{Diff}_+(S^1)$. Are there any other 1-cocycles with the same properties and what are the corresponding geometrical problems? How about 2-cocycles? These questions will be discussed in the present chapter.

[1] Sophie Germain.

As in the previous chapter, we consider here the one-dimensional case. The "circle" should be understood as the "projective line"; however, we use a more traditional notation S^1 instead of $\mathbb{RP}^1$.

3.1 Transvectants

An algebraic structure usually means a product, that is, a bilinear operation on a vector space. In this section we study bilinear PGL(2, $\mathbb{R}$)-invariant differential operators on the space of tensor densities on S^1. More precisely, we classify PGL(2, $\mathbb{R}$)-invariant differential operators from $\mathcal{F}_\lambda(S^1) \otimes \mathcal{F}_\mu(S^1)$ to $\mathcal{F}_\nu(S^1)$.

Main theorem

For the sake of brevity, let us formulate the result for generic λ and μ.

Theorem 3.1.1 *(i) For every λ, μ and integer $m \geq 0$, there exists a* PGL(2, $\mathbb{R}$)*-invariant bilinear differential operator of order m from $\mathcal{F}_\lambda \otimes \mathcal{F}_\mu$ to $\mathcal{F}_{\lambda+\mu+m}$ given by the formula*

$$J_m^{\lambda,\mu}(\phi, \psi) = \sum_{i+j=m} (-1)^j \binom{2\lambda + m - 1}{j} \binom{2\mu + m - 1}{i} \phi^{(i)} \psi^{(j)}, \tag{3.1}$$

where ϕ stands for $\phi(x)(\mathrm{d}x)^\lambda$ and $\phi^{(i)}$ for $(\mathrm{d}^i\phi(x)/\mathrm{d}x^i)(\mathrm{d}x)^{\lambda+i}$ and likewise for ψ.
(ii) If either λ or μ do not belong to the set $\{0, -\frac{1}{2}, -1, \ldots, -\frac{m-1}{2}\}$ then the operator $J_m^{\lambda,\mu}$ is the unique (up to a constant) bilinear PGL(2, $\mathbb{R}$)*-invariant differential operator from $\mathcal{F}_\lambda \otimes \mathcal{F}_\mu$ to $\mathcal{F}_{\lambda+\mu+m}$.*
(iii) If $\nu \neq \lambda + \mu + m$, then there are no PGL(2, $\mathbb{R}$)*-invariant differential operators from $\mathcal{F}_\lambda \otimes \mathcal{F}_\mu$ to $\mathcal{F}_\nu$.*

The proof is similar to that of Theorem 2.1.2 and we do not dwell on it. The operators $J_m^{\lambda,\mu}$ are called *transvectants*.

Examples of transvectants

The first-order transvectant $J_1^{\lambda,\mu}$ is invariant with respect to the full group Diff($\mathbb{RP}^1$). It is known as the (one-dimensional) Schouten bracket. Let us see how the second-order transvectants behave under arbitrary diffeomorphisms.

Exercise 3.1.2 Apply a diffeomorphism $f \in \mathrm{Diff}(\mathbb{RP}^1)$ to $J_2^{\lambda,\mu}$ and prove that

$$\mathrm{T}_f^{\lambda+\mu+2} \circ J_2^{\lambda,\mu}\left(\mathrm{T}_{f^{-1}}^{\lambda}\phi, \mathrm{T}_{f^{-1}}^{\mu}\psi\right) = J_2^{\lambda,\mu}(\phi, \psi) + 4\lambda\mu(\lambda+\mu+1)\, S(f)\, \phi\psi$$

where $S(f)$ is the Schwarzian derivative.

It follows that $J_2^{\lambda,\mu}$ is $\mathrm{Diff}(\mathbb{RP}^1)$-invariant if and only if $\lambda\mu(\lambda+\mu+1) = 0$.

Of transvectants of higher order we mention $J_3^{-2/3,-2/3}$. For two tensor densities $\phi = \phi(x)\,(\mathrm{d}x)^{-2/3}$ and $\psi = \psi(x)\,(\mathrm{d}x)^{-2/3}$, this operator acts as follows:

$$\phi \otimes \psi \mapsto \left(2\begin{vmatrix}\phi(x) & \psi(x)\\ \phi(x)''' & \psi(x)'''\end{vmatrix} + 3\begin{vmatrix}\phi(x)' & \psi(x)'\\ \phi(x)'' & \psi(x)''\end{vmatrix}\right)(\mathrm{d}x)^{5/2}$$

Exercise 3.1.3 Check that this operator is $\mathrm{Diff}(\mathbb{RP}^1)$-invariant.

This remarkable operator was discovered by P. Grozman.

Transvectants and symplectic geometry

Perhaps the best way to understand the operators (3.1) is to rewrite them in terms of symplectic geometry; the reader is encouraged to consult Appendix A.2.

Consider the plane $\mathbb{R}^2$ with linear coordinates (p, q) and the standard symplectic form $\omega = \mathrm{d}p \wedge \mathrm{d}q$. The symmetry group in this case is the group $\mathrm{Sp}(2,\mathbb{R})$ of linear symplectic transformations. Note that this group is isomorphic to $\mathrm{SL}(2,\mathbb{R})$. We will describe bilinear differential operators

$$B : C^\infty(\mathbb{R}^2) \otimes C^\infty(\mathbb{R}^2) \to C^\infty(\mathbb{R}^2),$$

invariant with respect to the action of $\mathrm{Sp}(2,\mathbb{R})$. As usual, it is easier to deal with the corresponding action of the Lie algebra $\mathrm{sp}(2,\mathbb{R})$. This action is generated by the Hamiltonian functions

$$\{p^2, \quad pq, \quad q^2\}.$$

Example 3.1.4 The simplest $\mathrm{Sp}(2,\mathbb{R})$-invariant operators are the product of functions $F \otimes G \mapsto FG$, and the Poisson bracket

$$\{F, G\} = F_p G_q - F_q G_p.$$

All bilinear $\mathrm{Sp}(2,\mathbb{R})$-invariant differential operators are, in fact, "iterations" of the Poisson bracket. In order to construct such iterations, let us notice that the Poisson bracket is the composition of the operator P on $C^\infty(\mathbb{R}^2)^{\otimes 2}$

$$P(F, G) = F_p \otimes G_q - F_q \otimes G_p \tag{3.2}$$

and the natural projection $\mathrm{Tr} : C^\infty(\mathbb{R}^2)^{\otimes 2} \to C^\infty(\mathbb{R}^2)$. For every m, define a bilinear differential operator of order $2m$ by

$$B_m := \frac{1}{m!}\,\mathrm{Tr} \circ P^m. \tag{3.3}$$

The explicit formula of the operator B_m is much simpler than that of the transvectants. The operators B_m are written in terms of "differential binomial".

Exercise 3.1.5 Calculate two further examples

$$B_2(F, G) = \tfrac{1}{2}\left(F_{pp}G_{qq} - 2\,F_{pq}G_{pq} + F_{qq}G_{pp}\right),$$
$$B_3(F, G) = \tfrac{1}{6}\left(F_{ppp}G_{qqq} - 3\,F_{ppq}G_{pqq} + 3\,F_{pqq}G_{ppq} - F_{qqq}G_{ppp}\right).$$

The following statement is an analog of Theorem 3.1.1. Its proof is similar to that of Theorem 2.1.2 and will also be omitted.

Proposition 3.1.6 *(i) The operator B_m is* Sp(2, $\mathbb{R}$)*-invariant for every m.*
(ii) There are no other bilinear Sp(2, $\mathbb{R}$)*-invariant differential operators on* $C^\infty(\mathbb{R}^2)$.

The group Sp(2, $\mathbb{R}$) is the double covering of PGL(2, $\mathbb{R}$), and it is natural to compare their invariants. It turns out that the transvectants (3.1) coincide with the iterated Poisson brackets (3.3). Let us identify the space of tensor densities $\mathcal{F}_\lambda(\mathbb{RP}^1)$ and the space of functions on $\mathbb{R}^2 \setminus \{0\}$ (with singularities at the origin) homogeneous of degree -2λ by the formula

$$\phi = \phi(x)\,(\mathrm{d}x)^\lambda \longmapsto F_\phi(p, q) = p^{-2\lambda}\,\phi\left(\frac{q}{p}\right) \tag{3.4}$$

where the affine coordinate on $\mathbb{RP}^1$ is chosen as $x = q/p$.

Theorem 3.1.7 *One has*

$$B_m(F_\phi, F_\psi) = F_{J_m^{\lambda,\mu}(\phi,\psi)}. \tag{3.5}$$

Proof Tensor densities of degree λ on $\mathbb{RP}^1$ can be viewed as homogeneous functions on the cotangent bundle $T^*\mathbb{RP}^1$ of degree $-\lambda$. In local coordinates (x, ξ) on $T^*\mathbb{RP}^1$, the correspondence between tensor densities and functions is as follows:

$$\phi = \phi(x)\,(\mathrm{d}x)^\lambda \longmapsto \Phi_\phi(x, \xi) = \phi(x)\,\xi^{-\lambda}.$$

The group PGL(2, $\mathbb{R}$) and the Lie algebra sl(2, $\mathbb{R}$) naturally act on $T^*\mathbb{RP}^1$. The action of sl(2, $\mathbb{R}$) is generated by the Hamiltonian functions

$$\{x, \ x\xi, \ x\xi^2\}.$$

The cotangent bundle with the zero section removed, $T^*\mathbb{RP}^1 \setminus \mathbb{RP}^1$, consists of two open cylinders. Denote by $T^*\mathbb{RP}^1_+$ the upper cylinder and consider the diffeomorphism $\mathbb{R}^2 \setminus \{0\} \to T^*\mathbb{RP}^1_+$ given by

$$(x, \xi) = \left(\frac{p^2}{2}, \ \frac{q}{p}\right). \tag{3.6}$$

Obviously the diffeomorphism (3.6) intertwines the actions of sp(2, $\mathbb{R}$) and sl(2, $\mathbb{R}$). By uniqueness (see Theorem 3.1.1 and Proposition 3.1.6), the sp(2, $\mathbb{R}$)-invariant operator B_m and the sl(2, $\mathbb{R}$)-invariant operator J_m have to be proportional after the identification (3.4).

To calculate precisely the coefficient of proportionality, it suffices to pick two arbitrary densities, say $\phi = (\mathrm{d}x)^\lambda$ and $\psi = x^m\,(\mathrm{d}x)^\mu$.

Exercise 3.1.8 Check that $J_m^{\lambda,\mu}(\phi, \psi) = (-1)^m\,(2\lambda)_m\,(\mathrm{d}x)^{\lambda+\mu+m}$ where $(a)_m = a(a+1)\cdots(a+m-1)$.

The corresponding functions on $\mathbb{R}^2$ are given by formula (3.4). One obtains $F_\phi = p^{-2\lambda}$ and $F_\psi = p^{-2\mu-m}\,q^m$.

Exercise 3.1.9 Check that $B_m(F_\phi, F_\psi) = (-1)^m\,(2\lambda)_m\,p^{-2(\lambda+\mu+m)}$.

Theorem 3.1.7 is proved. □

This symplectic viewpoint will be useful for multi-dimensional generalizations of transvectants.

Comment

Surprisingly enough, bilinear PGL(2, $\mathbb{R}$)-invariant differential operators were discovered earlier than the linear ones. The operators (3.1) were found by Gordan [86] in 1885 in the framework of invariant theory. Transvectants have been rediscovered more than once: by R. Rankin [177] in 1956, H. Cohen [43] in 1975 ("Rankin–Cohen brackets") and by S. Janson and J. Peetre in 1987 [103]; see also [157]. Theorem 3.1.7 was proved in [167].

Binary Diff(M)-invariant differential operators on an arbitrary manifold M are classified by P. Grozman [87]. The list of these operators includes well-known classical ones, such as Schouten brackets, Nijenhuis brackets, etc. The most remarkable operator in the Grozman list is $J_3^{-2/3,-2/3}$; this operator has no analogs in the multi-dimensional case.

3.2 First cohomology of Diff(S^1) with coefficients in differential operators

In this section we consider the first cohomology of Diff(S^1) and Vect(S^1) with coefficients in the space of differential operators $\mathcal{D}_{\lambda,\mu}(S^1)$ from $\mathcal{F}_\lambda(S^1)$ to $\mathcal{F}_\mu(S^1)$ (see Section 1.5). More specifically, we will be interested in the PGL(2, $\mathbb{R}$)-*relative* cohomology. Recall our assumption that all the cocycles on Diff(S^1) are given by differentiable maps.

We "rediscover" the classic Schwarzian derivative, as well as two other non-trivial cocycles on Diff(S^1) vanishing on PGL(2, $\mathbb{R}$). These cocycles are higher analogs of the Schwarzian derivative. These results allow us to study the Diff(S^1)-module $\mathcal{D}_{\lambda,\mu}(S^1)$ as a deformation of the module of tensor densities.

Schwarzian, as a cocycle on Diff(S^1), revisited

Consider 1-cocycles on Diff(S^1) with values in the space of differential operators:

$$C : \mathrm{Diff}(S^1) \to \mathcal{D}_{\lambda,\mu}(S^1). \tag{3.7}$$

The cocycle relation reads

$$C(f \circ g) = \mathrm{T}^\mu_f \circ C(g) \circ \mathrm{T}^\lambda_{f^{-1}} + C(f) \tag{3.8}$$

where T^λ is the Diff(S^1)-action on λ-densities (1.27) and the first summand on the right hand side is the Diff(S^1)-action (2.1) on $\mathcal{D}_{\lambda,\mu}(S^1)$. We will classify the cocycles (3.7) satisfying an additional condition: the restriction to the subgroup PGL(2, $\mathbb{R}$) is identically zero.

The classic Schwarzian derivative provides an example of such cocycle.

Example 3.2.1 Define a map S_λ : Diff(S^1) $\to \mathcal{D}_{\lambda,\lambda+2}(S^1)$ as the zero-order operator of multiplication by the Schwarzian derivative, namely, for $f \in$ Diff(S^1), set

$$S_\lambda(f^{-1}) : \phi \longmapsto S(f)\,\phi \tag{3.9}$$

where $\phi \in \mathcal{F}_\lambda(S^1)$. Since $S(f)$ is a quadratic differential, the right hand side belongs to $\mathcal{F}_{\lambda+2}(S^1)$. The Schwarzian derivative is a 1-cocycle on Diff(S^1)

with values in the space $\mathcal{F}_2(S^1)$ of quadratic differentials; see Section 1.5. This implies the cocycle relation for the map (3.9).

Three cocycles generalizing Schwarzian derivative

As in Section 2.5, let us use the notation $\delta = \mu - \lambda$. We will prove the following classification theorem.

Theorem 3.2.2 *For*

$$\delta \in \{2, 3, 4\}, \tag{3.10}$$

with arbitrary $\lambda \neq -\frac{1}{2}, -1$ *and* $-\frac{3}{2}$, *respectively, and for*

$$(\lambda, \mu) \in \{(-4, 1), (0, 5)\}, \tag{3.11}$$

there exists a unique (up to a constant) non-trivial 1-cocycle (3.7) vanishing on the subgroup PGL(2, $\mathbb{R}$). *Otherwise, there are no such non-trivial cocycles.*

In other words, there are three one-parameter families and two exceptional cocycles.

Let us give the explicit formulæ. If $\delta = 2$ then the cocycle is (3.9). If $\delta = 3, 4$ then the cocycles are as follows:

$$U_\lambda(f^{-1}) = S(f)\,\frac{\mathrm{d}}{\mathrm{d}x} - \frac{\lambda}{2}\,S(f)',$$

$$\begin{aligned} V_\lambda(f^{-1}) &= S(f)\,\frac{\mathrm{d}^2}{\mathrm{d}x^2} - \frac{2\lambda+1}{2}\,S(f)'\,\frac{\mathrm{d}}{\mathrm{d}x} \\ &\quad + \frac{\lambda(2\lambda+1)}{10}\,S(f)'' - \frac{\lambda(\lambda+3)}{5}\,S(f)^2, \end{aligned} \tag{3.12}$$

respectively. We do not give here explicit expressions for the exceptional cocycles corresponding to the values (3.11).

Exercise 3.2.3 Check the cocycle relation for U_λ and V_λ.

Theorem 3.2.2 can be reformulated in terms of PGL(2, $\mathbb{R}$)-relative cohomology of Diff(S^1).

Corollary 3.2.4 *There are three one-parameter families of* (λ, μ)*:*

$$(\delta = 2,\ \lambda \neq -\tfrac{1}{2}), \qquad (\delta = 3,\ \lambda \neq -1), \qquad (\delta = 4,\ \lambda \neq -\tfrac{3}{2}),$$

and two isolated values, $(\lambda, \mu) = (-4, 1), (0, 5)$, *for which*

$$H^1(\mathrm{Diff}(S^1), \mathrm{PGL}(2, \mathbb{R}); \mathcal{D}_{\lambda,\mu}(S^1)) = \mathbb{R}.$$

Otherwise, this cohomology is trivial.

Recall that there exist two more cocycles on $\mathrm{Diff}(S^1)$ with values in tensor densities $f^{-1} \mapsto \ln f'(x)$ and $f^{-1} \mapsto (f''(x)/f'(x))\mathrm{d}x$ with values in $\mathcal{F}_0(S^1)$ and $\mathcal{F}_1(S^1)$, respectively; see Theorem 1.5.5. The following exercise shows the exceptional role of the Schwarzian.

Exercise 3.2.5 Check that the 1-cocycles with values in $\mathcal{D}_{\lambda,\lambda}(S^1)$ and $\mathcal{D}_{\lambda,\lambda+1}(S^1)$, defined via multiplication by these 1-cocycles, are trivial for all λ, except $\lambda = 0$.

We now start the proof of Theorem 3.2.2.

Relation to PGL(2, $\mathbb{R}$)-invariant differential operators

Recall that a 1-cocycle (3.7) is a coboundary if there exists $B \in \mathcal{D}_{\lambda,\mu}$ such that

$$C(f) = \mathrm{T}^{\mu}_{f} \circ B \circ \mathrm{T}^{\lambda}_{f^{-1}} - B \tag{3.13}$$

for every $f \in \mathrm{Diff}(S^1)$; cf. Section 1.5. One then writes $C = d(B)$.

If a coboundary (3.13) vanishes on PGL(2, $\mathbb{R}$) then the operator B is PGL(2, $\mathbb{R}$)-invariant. We already classified such operators in Section 2.1; see Theorem 2.1.2. It follows that every coboundary vanishing on PGL(2, $\mathbb{R}$) is proportional to $d\,(\mathrm{D}_k)$ where D_k is the PGL(2, $\mathbb{R}$)-invariant differential operator (2.3).

Exercise 3.2.6 Check that the cocycles $S_{-1/2}$, U_{-1} and $V_{-3/2}$ are coboundaries:

$$S_{-1/2} = d\,(\mathrm{D}_2), \qquad U_{-1} = d\,(\mathrm{D}_3), \qquad V_{-3/2} = d\,(\mathrm{D}_4).$$

For other values of λ the cocycles S_λ, U_λ and V_λ are not trivial.

Relation to transvectants: cohomology of Vect(S^1)

We will now classify infinitesimal analogs of the cocycles (3.7).

Consider the Lie algebra cocycles:

$$c : \mathrm{Vect}(S^1) \to \mathcal{D}_{\lambda,\mu}(S^1) \tag{3.14}$$

vanishing on the subalgebra sl(2, $\mathbb{R}$). The 1-cocycle relation reads:

$$c([X,Y]) = L^{\mu}_X \circ c(Y) - c(Y) \circ L^{\lambda}_X - L^{\mu}_Y \circ c(X) + c(X) \circ L^{\lambda}_Y \qquad (3.15)$$

for every $X, Y \in \mathrm{Vect}(S^1)$. This is just the infinitesimal version of (3.8).

Let us use a convenient property of 1-cocycles: a 1-cocycle, vanishing on a Lie subalgebra, is invariant with respect to this subalgebra. In our case we have the following consequence.

Lemma 3.2.7 *Given a 1-cocycle (3.14), vanishing on* sl(2, $\mathbb{R}$)*, the bilinear differential operator* $\mathrm{J} : \mathrm{Vect}(S^1) \otimes \mathcal{F}_\lambda \to \mathcal{F}_\mu$*, defined by*

$$\mathrm{J}(X, \phi) = c(X)(\phi), \qquad (3.16)$$

is sl(2, $\mathbb{R}$)*-invariant.*

Proof Consider $Y \in \mathrm{sl}(2,\mathbb{R})$. Since $c(Y) = 0$, one rewrites the condition (3.15) as

$$L^{\mu}_Y(\mathrm{J}(X,\phi)) - \mathrm{J}([Y,X],\phi) - \mathrm{J}(X, L^{\lambda}_Y(\phi)) = 0,$$

and the left hand side is precisely $L_Y(\mathrm{J})$, applied to (X, ϕ). □

One can now use Theorem 3.1.1 on classification of bilinear sl(2, $\mathbb{R}$)-invariant differential operators. It follows that the bilinear map (3.16) associated with a 1-cocycle (3.14) is necessarily proportional to a transvectant

$$\mathrm{J}(X,\phi) = c\, J^{-1,\lambda}_m(X,\phi)$$

where $\lambda = m - 1$ and c is a constant. It remains to check, one by one, which transvectants indeed define 1-cocycles on $\mathrm{Vect}(S^1)$.

Exercise 3.2.8 Check that

(a) $J^{-1,\lambda}_1(X,\phi) = L_X\phi$ and $J^{-1,\lambda}_2(X,\phi) = 0$ for all $X \in \mathrm{Vect}(S^1)$;
(b) up to a multiple,

$$J^{-1,\lambda}_3(X,\phi) = X'''\phi,$$

$$J^{-1,\lambda}_4(X,\phi) = X'''\phi' - \tfrac{\lambda}{2}\, X^{IV}\phi$$

$$J^{-1,\lambda}_5(X,\phi) = X'''\phi'' - \tfrac{2\lambda+1}{2}\, X^{IV}\phi' + \tfrac{\lambda(2\lambda+1)}{10}\, X^{V}\phi,$$

and these maps define 1-cocycles on $\mathrm{Vect}(S^1)$, vanishing on sl(2, $\mathbb{R}$);
(c) the map $J^{-1,\lambda}_6$ defines a 1-cocycle if and only if $\lambda = -4, 0, -2$, and in the case $\lambda = -2$, it is a coboundary, namely, $d(\mathrm{D}_5)$;

(d) the map $J_m^{-1,\lambda}$ with $m > 6$ and $\lambda \neq 1 - m/2$ never defines a 1-cocycle on Vect(S^1).

We obtain the following result:

$$H^1(\mathrm{Vect}(S^1), \mathrm{sl}(2, \mathbb{R}); \mathcal{D}_{\lambda,\mu}(S^1)) = \begin{cases} \mathbb{R}, & (\lambda, \mu) \text{ as in (3.10), (3.11)} \\ 0, & \text{otherwise,} \end{cases}$$

which is an infinitesimal analog of Theorem 3.2.2.

Proof of Theorem 3.2.2

A differentiable 1-cocycle on a Lie group G determines a 1-cocycle on the respective Lie algebra $\mathfrak{g}$:

$$c(X) = \frac{\mathrm{d}}{\mathrm{d}t} C(\exp tX)\Big|_{t=0}, \qquad X \in \mathfrak{g}; \tag{3.17}$$

see Appendix A.4.

In our case, we have just classified non-trivial cocycles on Vect(S^1), vanishing on sl(2, $\mathbb{R}$). We have also seen that these cocycles integrate to cocycles on Diff(S^1); cf. explicit formulæ (3.12). Since all the coboundaries are of the form $d(\mathrm{D}_k)$, the obtained cocycles are non-trivial and linearly independent except for the cases $\lambda = -1/2, -1$ and $-3/2$.

It remains to show that there are no other non-trivial cocycles on Diff(S^1) vanishing on sl(2, $\mathbb{R}$) than those of Theorem 3.2.2.

Lemma 3.2.9 *A non-zero differentiable 1-cocycle on G corresponds to a non-zero 1-cocycle on $\mathfrak{g}$.*

Proof Assume C is a 1-cocycle on G such that the 1-cocycle on $\mathfrak{g}$ given by formula (3.17) vanishes, and let us show that C is identically zero. We claim that if the differential of C vanishes at the unit element then it vanishes at every other point of G. Indeed, given a curve $g_t \in G$ with g_0 the unit element, then $\mathrm{d}\,C(g_t)/\mathrm{d}t = 0$ at $t = 0$. For every $g \in G$, one has, by the cocycle property,

$$\frac{\mathrm{d}}{\mathrm{d}t} C(g\, g_t)\Big|_{t=0} = \frac{\mathrm{d}}{\mathrm{d}t} (g\, C(g_t) + C(g))\Big|_{t=0} = 0.$$

The curve gg_t is an arbitrary curve at point $g \in G$ and we conclude that the differential of C vanishes on T_gG. Therefore $C = 0$. □

It follows from the above lemma that to every non-trivial cocycle on Diff(S^1) there corresponds a non-trivial cocycle on Vect(S^1).

Theorem 3.2.2 is proved.

Comment

Feigin and Fuchs calculated the first cohomology space of the Lie algebra of formal vector fields on $\mathbb{R}$ with coefficients in the module of differential operators (see [65] and also [72]). They did not study a similar problem in the case of diffeomorphism groups. Theorem 3.2.2 was proved in [33].

3.3 Application: geometry of differential operators on $\mathbb{RP}^1$

Geometry is the main subject of this book and the reappearance of this term in the title of this section might look strange. Recall that geometry is understood, in the sense of Klein, as a Lie group action on a manifold. Two groups play a special role in this book, namely the group of projective transformations and the full group of diffeomorphisms.

In this section we study the space of k-th order differential operators $\mathcal{D}^k_{\lambda,\mu}(S^1)$ as a module over the group Diff(S^1) and solve the problem of their classification. Our main tool is the canonical isomorphism between the space $\mathcal{D}^k_{\lambda,\mu}(S^1)$ and the corresponding space of symbols $\mathcal{S}_\delta(S^1)$ as PGL(2, $\mathbb{R}$)-modules, introduced in Section 2.5.

Formulating the classification problem

The classification problem is formulated as follows: for which values (λ, μ) and (λ', μ') is there an isomorphism of Diff(S^1)-modules

$$\mathcal{D}^k_{\lambda,\mu}(S^1) \cong \mathcal{D}^k_{\lambda',\mu'}(S^1)? \tag{3.18}$$

Example 3.3.1 We have already encountered such an isomorphism more than once, namely the conjugation:

$$* : \mathcal{D}_{\lambda,\mu}(S^1) \longrightarrow \mathcal{D}_{1-\mu,1-\lambda}(S^1).$$

The question is, therefore, whether there are other isomorphisms.

The "shift" of the degree

$$\delta = \mu - \lambda \tag{3.19}$$

is obviously an invariant, since the modules of symbols $\mathcal{S}^k_\delta(S^1)$ with different δ are not isomorphic to each other. Indeed, one has

$$\mathcal{S}_\delta(S^1) = \bigoplus_{\ell=0}^{\infty} \mathcal{F}_{\delta-\ell}(S^1).$$

We are thus led to classifying the modules $\mathcal{D}^k_{\lambda,\mu}(S^1)$ with fixed δ.

Results

It turns out that, for $k \leq 3$, almost all modules $\mathcal{D}^k_{\lambda,\mu}(S^1)$ with fixed δ are isomorphic.

Theorem 3.3.2 *If $k \leq 3$ then all the modules $\mathcal{D}^k_{\lambda,\mu}(S^1)$ with fixed δ are isomorphic to each other, except for the values of (λ, μ) given by the table:*

k	1	2	3
(λ, μ)	$(0, 1)$	$(0, \mu)$	$(0, \mu)$
		$(-\frac{1}{2}, \frac{3}{2})$	$(\lambda, 1-\lambda)$
			$(\lambda, \frac{4\lambda+1}{3\lambda+1})$
			$(\lambda, \lambda+2)$

(3.20)

and their adjoint ones: $\mathcal{D}^k_{1-\mu,1-\lambda}(S^1)$.

The values of (λ, μ) corresponding to the exceptional modules of third-order operators are given in Figure 3.1.

In the general case the result is different.

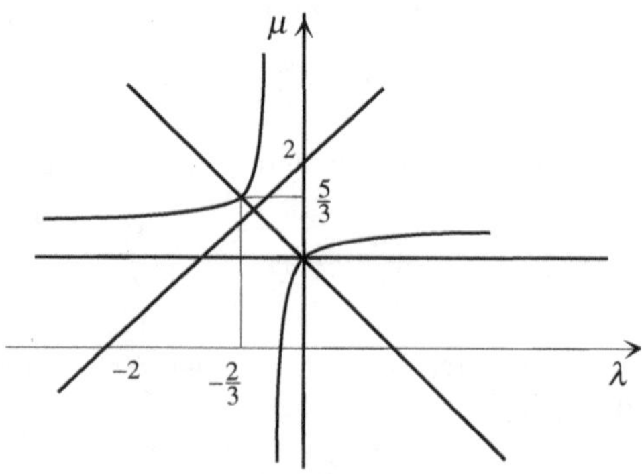

Fig. 3.1. Exceptional modules of third-order operators

Theorem 3.3.3 *If $k \geq 4$ then, up to a constant, the only isomorphism (3.18) is the conjugation.*

We now begin the proof of these theorems. The main ingredient is the isomorphism between the spaces of differential operators and symbols, viewed as PGL(2, $\mathbb{R}$)-modules.

Diff(S^1)-action in the canonical form

Using the projectively equivariant symbol map (2.21), we can rewrite the Diff(S^1)-action (2.1) in terms of the symbols, namely,

$$\begin{array}{ccc} \mathcal{D}_{\lambda,\mu} & \xrightarrow{\mathrm{T}_f^{\lambda,\mu}} & \mathcal{D}_{\lambda,\mu} \\ {\scriptstyle \sigma_{\lambda,\mu}}\downarrow & & \downarrow{\scriptstyle \sigma_{\lambda,\mu}} \\ \mathcal{S}_\delta & \xrightarrow{\sigma_{\lambda,\mu}\circ \mathrm{T}_f^{\lambda,\mu}\circ\sigma_{\lambda,\mu}^{-1}} & \mathcal{S}_\delta \end{array} \qquad (3.21)$$

Let us compare the action $\sigma_{\lambda,\mu} \circ \mathrm{T}_f^{\lambda,\mu} \circ \sigma_{\lambda,\mu}^{-1}$ with the standard Diff(S^1)-action on $\mathcal{S}_\delta(S^1)$. Every $\Phi \in \mathcal{S}_\delta(S^1)$ is of the form:

$$\Phi = \sum_{\ell=0}^{k} \Phi_\ell(x)(dx)^{\delta-\ell}$$

for some k. Let

$$f(\Phi) = \sigma_{\lambda,\mu} \circ \mathrm{T}_f^{\lambda,\mu} \circ \sigma_{\lambda,\mu}^{-1}(\Phi).$$

It follows from Theorem 3.2.2 (the "uniqueness" part) that the higher-order terms of $f(\Phi)$ are necessarily of the form

$$\begin{aligned} f(\Phi)_k &= \mathrm{T}_f^{\delta-k}(\Phi_k), \\ f(\Phi)_{k-1} &= \mathrm{T}_f^{\delta-k+1}(\Phi_{k-1}), \\ f(\Phi)_{k-2} &= \mathrm{T}_f^{\delta-k+2}(\Phi_{k-2}) + \beta_2^k\, S_{\delta-k}(f)(\Phi_k), \\ f(\Phi)_{k-3} &= \mathrm{T}_f^{\delta-k+3}(\Phi_{k-3}) + \beta_3^k\, U_{\delta-k}(f)(\Phi_k) \\ &\quad + \beta_2^{k-1}\, S_{\delta-k+1}(f)(\Phi_{k-1}), \\ f(\Phi)_{k-4} &= \mathrm{T}_f^{\delta-k+4}(\Phi_{k-4}) + \beta_4^k\, V_{\delta-k}(f)(\Phi_k) \\ &\quad + \beta_3^{k-1}\, U_{\delta-k+1}(f)(\Phi_{k-1}) + \beta_2^{k-2}\, S_{\delta-k+2}(f)(\Phi_{k-2}) \end{aligned} \qquad (3.22)$$

where T^λ is the standard Diff(S^1)-action on the space of tensor densities, S, U

and V are the cocycles (3.9) and (3.12), and β_j^k are coefficients depending on λ and μ.

An algebraic meaning of the above expression is that the module $\mathcal{D}_{\lambda,\mu}(S^1)$ can be viewed as a *deformation* of the module $\mathcal{S}_\delta(S^1)$. The cocycles S, U and V describe its infinitesimal part while β_j^k are the parameters of the deformation.

Computing the parameters β_j^k

In order to obtain the classification results, we need some information about the coefficients in (3.22).

Exercise 3.3.4 Check that

$$\beta_2^4(\lambda,\mu) = -\frac{(6\lambda+4)\delta+6\lambda^2-6\lambda-5}{2\delta-7}$$

$$\beta_1^3(\lambda,\mu) = -\frac{(3\lambda+1)\delta+3\lambda^2-3\lambda-1}{2\delta-5}$$

$$\beta_1^4(\lambda,\mu) = \frac{(\delta+2\lambda-1)\,((4\lambda+1)\delta-4\lambda(\lambda-1))}{(\delta-2)(\delta-3)(\delta-4)}$$

$$\beta_0^2(\lambda,\mu) = -\frac{\lambda(\delta+\lambda-1)}{2\delta-3}$$

$$\beta_0^3(\lambda,\mu) = \frac{\lambda(\delta+\lambda-1)(\delta+2\lambda-1)}{(\delta-1)(\delta-2)(\delta-3)}$$

$$\beta_0^4(\lambda,\mu) = -\frac{\lambda(\delta+\lambda-1)(4\delta^2+12\lambda\delta-12\delta+12\lambda^2-12\lambda+11)}{(\delta-1)(2\delta-3)(2\delta-5)(2\delta-7)(\delta-4)}$$

Hint. It is easier to compute the Vect(S^1)-action; the formula (3.22) remains the same but the cocycles S_λ, U_λ and V_λ should be replaced by the corresponding 1-cocycles on Vect(S^1), i.e., by the transvectants J_3, J_4 and J_5. It follows from the definition (3.21) that the general formula for the coefficients is

$$\beta_j^k = \frac{(-1)^j}{\binom{j+2\delta-2k-2}{3}}\left((\delta-k)\,C_j^k - \sum_{i=1}^{j}\left(\lambda\binom{k}{i}+\binom{k}{i+1}\right)C_{j-i}^{k-i}\right)$$

where C_m^ℓ are as in (2.22). Specifying k and s, one then obtains the above formulæ (computer-assisted symbolic computation is highly recommended).

Proof of Theorem 3.3.2

Consider the linear map $T : \mathcal{D}^3_{\lambda,\mu} \to \mathcal{D}^3_{\lambda',\mu'}$ defined, in terms of the projectively equivariant symbol, by

$$T(\Phi_3\,\xi^3 + \Phi_2\,\xi^2 + \Phi_1\,\xi + \Phi_0) = \Phi_3\,\xi^3 + \frac{\beta'^3_0\beta^2_0}{\beta^3_0\beta'^2_0}\,\Phi_2\,\xi^2 + \frac{\beta'^3_1}{\beta^3_1}\,\Phi_1\,\xi^1 + \frac{\beta'^3_0}{\beta^3_0}\Phi_0$$

This map intertwines the actions (3.22) with (λ, μ) and (λ', μ'). This proves Theorem 3.3.2 for non-resonant values of δ (i.e., for $\delta \neq -1, -\frac{3}{2}, -2, -\frac{5}{2}, -3$) since we used the projectively equivariant symbol map in the construction of the isomorphism.

Exercise 3.3.5 Prove that the isomorphism T makes sense for the resonant values of δ as well.

Hint. Rewrite the formula of the map T in terms of the coefficients of differential operators.

This completes the proof of Theorem 3.3.2. □

Proof of Theorem 3.3.3

For the sake of simplicity, we give here the proof for generic, non-resonant values of δ.

Consider an isomorphism $T : \mathcal{D}^k_{\lambda,\mu}(S^1) \to \mathcal{D}^k_{\lambda',\mu'}(S^1)$ with $k \geq 4$. Since T is an isomorphism of $\mathrm{Diff}(S^1)$-modules, it is also an isomorphism of $\mathrm{PGL}(2, \mathbb{R})$-modules. The uniqueness of the projectively equivariant symbol map shows that the linear map $\sigma \circ T \circ \sigma^{-1}$ on $\mathcal{S}^k_\delta(S^1)$ is multiplication by a constant on each homogeneous component:

$$T(P_4\xi^4 + \cdots + P_0) = P_4\xi^4 + \tau_3 P_3\xi^3 + \cdots + \tau_0 P_0,$$

with τ_i depending on $\lambda, \mu, \lambda', \mu'$. This map intertwines two $\mathrm{Diff}(S^1)$-actions (3.22) with (λ, μ) and (λ', μ') if and only if

$$\begin{aligned} \tau_4\beta'^4_2 &= \tau_2\beta^4_2, & \tau_4\beta'^4_1 &= \tau_1\beta^4_1 \\ \tau_4\beta'^4_0 &= \tau_0\beta^4_0, & \tau_3\beta'^3_1 &= \tau_1\beta^3_1 \\ \tau_3\beta'^3_0 &= \tau_0\beta^3_0, & \tau_2\beta'^2_0 &= \tau_0\beta^2_0. \end{aligned}$$

This system has a solution only for $\lambda = \lambda'$ or $\lambda + \mu' = -1$. The first isomorphism is tautological, the second one is the conjugation. This proves Theorem 3.3.3 for non-resonant δ. □

Comment

Examining the particular values of (λ, μ) from Theorem 3.3.2, one finds interesting modules of differential operators, already encountered in this book. For instance, the module $\mathcal{D}_{-1/2,3/2}(S^1)$ is precisely that related to the projective structures on S^1 and the Sturm–Liouville operators. Another interesting module is $\mathcal{D}^3_{-2/3,5/3}(S^1)$ (see Figure 3.1), related to the Grozman operator; see Exercise 3.1.2. The geometric meaning of this module is unclear yet.

Theorems 3.3.2 and 3.3.3 were proved in [75] for the particular case $\delta = 0$ and in [74] for the general case; see also [76].

3.4 Algebra of tensor densities on S^1

The space $\mathcal{F}(S^1)$ of all tensor densities on S^1 is a beautiful Poisson algebra. This space also has an infinite number of higher-order bilinear PGL(2, $\mathbb{R}$)-invariant operations, the transvectants; see Section 3.1.

In this section we explain how all these operations are related to each other. We will describe a rich family of algebraic structures on $\mathcal{F}(S^1)$ that can be viewed as a deformation of the natural product of tensor densities. These algebraic structures are based on the transvectants.

Poisson algebra of tensor densities

A *Poisson algebra* is a vector space $\mathfrak{A}$ equipped with a commutative associative product and a Lie algebra commutator denoted $\{\ ,\ \}$; these two algebra structures are related by the Leibniz identity

$$\{a, bc\} = \{a, b\}\, c + b\, \{a, c\}$$

for all $a, b, c \in \mathfrak{A}$. This means that the operator ad_a is a derivation of the commutative algebra.

The space $\mathcal{F}(S^1)$ has a natural (that is, a Diff(S^1)-invariant) structure of a Poisson algebra with the commutative product given by the usual product of tensor densities

$$\phi(x)(\mathrm{d}x)^\lambda \otimes \psi(x)(\mathrm{d}x)^\mu \mapsto \phi(x)\,\psi(x)(\mathrm{d}x)^{\lambda+\mu}$$

and the Lie algebra commutator given by the *Schouten bracket*

$$\{\phi(x)(\mathrm{d}x)^\lambda, \psi(x)(\mathrm{d}x)^\mu\} = \big(\lambda\,\phi(x)\,\psi'(x) - \mu\,\psi'(x)\,\phi(x)\big)\ (\mathrm{d}x)^{\lambda+\mu+1}.$$

In our notation, the latter is the first transvectant $J_1^{\lambda,\mu}$; see formula (3.1).

Exercise 3.4.1 Check the Leibniz identity for the above operations.

We will also consider the space $\mathcal{S}(S^1)$ of symbols of differential operators; see Section 2.5. This space is a direct sum of the tensor density spaces:

$$\mathcal{S}(S^1) = \bigoplus_{\ell \leq 0} \mathcal{F}_\ell(S^1).$$

The space $\mathcal{S}(S^1)$ is a graded Poisson subalgebra of $\mathcal{F}(S^1)$.

Remark 3.4.2 The Lie algebra $\mathrm{Vect}(S^1)$ is a Lie subalgebra of $\mathcal{F}(S^1)$. Moreover, the Lie derivative of a tensor density is just the Poisson bracket:

$$L_X(\phi) = \{\phi, X\}.$$

Deformation of Poisson algebras: star-products

We introduce the notion of a *star-product* on a Poisson algebra. This operation is an associative (but not necessarily commutative) deformation of the commutative product still satisfying the Leibnitz identity.

A star-product on a Poisson algebra $\mathfrak{A}$ is given by a series

$$a * b = a\,b + \frac{t}{2}\,\{a, b\} + \sum_{k=2}^{\infty} \frac{t^k}{2^k}\,B_k(a, b) \tag{3.23}$$

where t is a (formal) parameter and $B_k : \mathfrak{A}^{\otimes 2} \to \mathfrak{A}$ are bilinear maps, satisfying the following *associativity* property. If one extends (by linearity) the product (3.23) to the space of formal series $\mathfrak{A}[[t]] = \mathfrak{A} \otimes \mathbb{R}[[t]]$, then

$$a * (b * c) = (a * b) * c. \tag{3.24}$$

Exercise 3.4.3 Check that relation (3.24) is always satisfied up to the first order.

Hint. This is equivalent to the Leibnitz identity.

Relation (3.24) is a strong restriction on the bilinear maps B_k. For a given Poisson algebra $\mathfrak{A}$, the existence of a star-product is, a-priori, not guaranteed.

The associative algebra $(\mathfrak{A}[[t]], *)$ naturally carries a Lie algebra structure, given by the *star-commutator*

$$[a, b]_* = \frac{a * b - b * a}{t} \tag{3.25}$$

which is a deformation of the Poisson bracket.

Example 3.4.4 The simplest and most famous example of a star-product is the so-called *Moyal product* on symplectic vector space $\mathbb{R}^{2n}$. In the case of

$\mathbb{R}^2$, this star-product is given by the bilinear operations (3.3). In the multi-dimensional case, the definition is analogous. The associativity condition can be easily checked.

The Moyal star-product on tensor densities

The identification (3.4) maps the algebra of tensor densities $\mathcal{F}(S^1)$ to the algebra of homogeneous functions on $\mathbb{R}^2 \setminus \{0\}$; the Schouten bracket corresponds to the standard Poisson bracket on $\mathbb{R}^2$. The pull-back of the Moyal product on $\mathbb{R}^2$ then defines a star-product on tensor densities which we will also call the Moyal product.

Proposition 3.4.5 *The explicit formula for the star-product on $\mathcal{F}(S^1)$ is as follows:*

$$\phi * \psi = \phi\,\psi + \frac{t}{2}\,\{\phi, \psi\} + \sum_{k=2}^{\infty} \frac{t^k}{2^k}\, J_k^{\lambda,\mu}(\phi, \psi) \tag{3.26}$$

where $\phi \in \mathcal{F}_\lambda(S^1)$, $\psi \in \mathcal{F}_\mu(S^1)$ and $J_k^{\lambda,\mu}$ are the transvectants (3.1).

Proof The transvectants $J_k^{\lambda,\mu}$ correspond, after the identification (3.4), to the terms B_k of the Moyal product given by (3.3) see Theorem 3.1.7. Formula (3.26) is therefore precisely formula (3.23), rewritten in terms of tensor densities. □

The star-product (3.26) is obviously PGL(2, $\mathbb{R}$)-invariant.

The Cohen–Manin–Zagier star-product

The algebra $\mathcal{F}(S^1)$ has a natural family of star-products. Consider the projectively invariant symbol map (see (2.21)–(2.22)) and the quantization map (2.24) in the particular case $\lambda = \mu = \nu$:

$$\sigma_\nu : \mathcal{D}_\nu(S^1) \to \mathcal{S}(S^1), \qquad \mathcal{Q}_\nu : \mathcal{S}(S^1) \to \mathcal{D}_\nu(S^1)$$

with $\nu \in \mathbb{R}$. Of course, one has $\mathcal{Q}_\nu = \sigma_\nu^{-1}$.

Consider two tensor densities $\phi \in \mathcal{F}_\lambda(S^1)$ and $\psi \in \mathcal{F}_\mu(S^1)$ and assume first that λ and μ are *non-positive integers*. In other words, we take $\phi, \psi \in \mathcal{S}(S^1)$. Define a new product by

$$\phi *_\nu \psi = \mathcal{Q}_\nu^{-1}\left(\mathcal{Q}_\nu(\phi)\, \mathcal{Q}_\nu(\psi)\right). \tag{3.27}$$

This operation is obviously associative since it comes from the associative product in the algebra of differential operators.

Theorem 3.4.6 *The explicit formula for the defined product on the space of tensor densities is*

$$\phi *_\nu \psi = \phi \psi + \frac{1}{2} \{\phi, \psi\} + \sum_{k=2}^{\infty} B_k^\nu(\lambda, \mu) J_k^{\lambda,\mu}(\phi, \psi) \tag{3.28}$$

where the coefficients $B_k^\nu(\lambda, \mu)$ are given by

$$B_k^\nu(\lambda, \mu) = \left(\frac{1}{2}\right)^k \sum_{j \geq 0} \binom{k}{2j} \frac{\binom{-\frac{1}{2}}{j}\binom{2\nu-\frac{3}{2}}{j}\binom{\frac{1}{2}-2\nu}{j}}{\binom{-2\lambda-\frac{1}{2}}{j}\binom{-2\mu-\frac{1}{2}}{j}\binom{k+2\lambda+2\mu-\frac{3}{2}}{j}} \tag{3.29}$$

We will not give here a complete proof of this theorem; see [44]. By construction, the product (3.27) is PGL(2, $\mathbb{R}$)-invariant; it thus has to be expressed in terms of the transvectants, which makes formula (3.28) obvious. The explicit expression for the coefficients $B_k^\nu(\lambda, \mu)$ can be determined by choosing particular tensor densities; cf. proof of Theorem 3.1.7.

Formulæ (3.28) and (3.29) make sense for arbitrary λ and μ (not necessarily integer) and therefore define a one-parameter family of star-products on the whole Poisson algebra of tensor densities $\mathcal{F}(S^1)$.

Remark 3.4.7 (a) The product (3.27) is a composition of two differential operators written in a PGL(2, $\mathbb{R}$)-invariant way. The same is true for the Moyal product (3.26) but the quantization map is different.

(b) One can add the formal parameter t to the product (3.27) in the usual way. For instance, one considers the linear map

$$\mathcal{I} : \mathcal{S}(S^1) \to \mathcal{S}(S^1)[[t]]$$

such that $\mathcal{I}|_{\mathcal{F}_{-k}(S^1)} = t^k \,\mathrm{Id}$. Then one replaces the quantization and the symbol maps by $\bar{\mathcal{Q}}_\nu = \mathcal{Q}_\nu \circ \mathcal{I}$ and $\bar{\sigma_\nu} = \bar{\mathcal{Q}_\nu}^{-1}$, respectively.

Comment

The notion of star-product plays an important role in mathematical physics. It is closely related to quantum mechanics and is the main ingredient of deformation quantization. The first example of star-products is the Moyal product; it appeared in the physics literature in the first half of the twentieth century.

Theorem 3.4.6 was proved in [44]. The star-product (3.26) was considered in [167].

3.5 Extensions of Vect(S^1) by the modules $\mathcal{F}_\lambda(S^1)$

The Lie algebra Vect(S^1) has a non-trivial central extension called the Virasoro algebra; see Section 1.6. Since the Virasoro algebra plays such an important role in mathematics and theoretical physics, it is natural to look for its analogs and generalizations.

In this section we consider a natural class of "non-central" extensions of Vect(S^1), namely extensions by the modules of tensor densities $\mathcal{F}_\lambda(S^1)$. We will be interested in the projectively invariant extensions which are given by projectively invariant 2-cocycles. The result is quite surprising: there exists a unique extension if and only if $\lambda = 5$ or 7. As in the preceding section, the transvectants constitute the main ingredient of our constructions.

We thus obtain two exceptional infinite-dimensional Lie algebras that can be considered analogs of the Virasoro algebra. These Lie algebras are, in fact, Lie subalgebras of the algebra $\mathcal{F}(S^1)$ with respect to the product (3.26). Their existence is due to remarkable properties of the transvectants J_7 and J_9.

Statement of the problem

In this section we consider extensions of Vect(S^1) by the modules $\mathcal{F}_\lambda(S^1)$

$$0 \to \mathcal{F}_\lambda(S^1) \to \mathfrak{g} \to \mathrm{Vect}(S^1) \to 0. \tag{3.30}$$

In other words, we look for a Lie algebra structure on the vector space $\mathfrak{g} = \mathrm{Vect}(S^1) \oplus \mathcal{F}_\lambda(S^1)$ given by products of the form

$$[(X, \phi), (Y, \psi)] = \big([X, Y], L^\lambda_X \psi - L^\lambda_Y \phi + c(X, Y)\big),$$

where $X = X(x)\,\mathrm{d}/\mathrm{d}x$ is a vector field and L^λ_X is the Lie derivative (1.28). The bilinear skew-symmetric map $c : \mathrm{Vect}(S^1) \oplus \mathrm{Vect}(S^1) \to \mathcal{F}_\lambda(S^1)$ has to satisfy the condition

$$c\,(X, [Y, Z]) + L^\lambda_X\, c\,(Y, Z) + \mathrm{cycle}_{(X,Y,Z)} = 0, \tag{3.31}$$

which guarantees that the above commutator satisfies the Jacobi identity. In other words, c is a 2-cocycle; cf. Section 1.6 and Appendix A.4.

If $c = 0$ then the Lie algebra $\mathfrak{g}$ is called a semi-direct product. The extension (3.30) is *non-trivial* if the Lie algebra $\mathfrak{g}$ is not isomorphic to a semi-direct product. The cocycle c in this case represents a non-trivial cohomology class of the second cohomology space $H^2(\mathrm{Vect}(S^1); \mathcal{F}_\lambda(S^1))$. A trivial 2-cocycle is also called 2-*coboundary*. Such a cocycle can be written as

$$c\,(X, Y) = L^\lambda_X\, \ell\,(Y) - L^\lambda_Y\, \ell\,(X) - \ell\,([X, Y]) \tag{3.32}$$

where $\ell : \mathrm{Vect}(S^1) \to \mathcal{F}_\lambda(S^1)$ is a linear map.

As in Section 3.2, we will be interested in sl(2, $\mathbb{R}$)-relative cohomology, that is, we assume

$$c(X,\ Y) = 0 \quad \text{if} \quad X \in \mathrm{sl}(2, \mathbb{R}). \tag{3.33}$$

We will, furthermore, consider only the cocycles given by differential operators. The relevant cohomology space is therefore

$$H^2_{\mathrm{diff}}(\mathrm{Vect}(S^1), \mathrm{sl}(2, \mathbb{R}); \mathcal{F}_\lambda(S^1)).$$

Exercise 3.5.1 Check that a 2-cocycle c satisfying property (3.33) is sl(2, $\mathbb{R}$)-invariant, that is,

$$c([X,\ Y],\ Z) + c(Y,\ [X,\ Z]) = L^\lambda_X\, c(Y,\ Z) \tag{3.34}$$

for every $X \in \mathrm{sl}(2, \mathbb{R})$.

Hint. Use the cocycle condition (3.31) along with (3.33).

Two exceptional cocycles on Vect(S^1)

For the sl(2, $\mathbb{R}$)-relative cohomology space, one obtains the following result.

Theorem 3.5.2 *One has*

$$H^2_{\mathrm{diff}}(\mathrm{Vect}(S^1), \mathrm{sl}(2, \mathbb{R}); \mathcal{F}_\lambda(S^1)) = \begin{cases} \mathbb{R}, & \lambda = 5, 7, \\ 0, & \text{otherwise}. \end{cases}$$

Proof We start with the construction of non-trivial 2-cocycles with values in $\mathcal{F}_5(S^1)$ and $\mathcal{F}_7(S^1)$. These cocycles are just the transvectants J_7 and J_9, respectively. Indeed, these transvectants, restricted to $\mathrm{Vect}(S^1) \cong \mathcal{F}_{-1}(S^1)$, define skew-symmetric bilinear maps

$$J_7 : \mathrm{Vect}(S^1)^{\otimes 2} \to \mathcal{F}_5(S^1), \qquad J_9 : \mathrm{Vect}(S^1)^{\otimes 2} \to \mathcal{F}_7(S^1)$$

where, to simplify the notation, we omit the upper indices: $J_k = J_k^{-1,-1}$.

Exercise 3.5.3 Check that, up to a multiple, one has

$$J_7(X,\ Y) = \begin{vmatrix} X''' & Y''' \\ X^{(IV)} & Y^{(IV)} \end{vmatrix} (\mathrm{d}x)^5$$

and

$$J_9(X,\ Y) = \left(2 \begin{vmatrix} X''' & Y''' \\ X^{(VI)} & Y^{(VI)} \end{vmatrix} - 9 \begin{vmatrix} X^{(IV)} & Y^{(IV)} \\ X^{(V)} & Y^{(V)} \end{vmatrix} \right) (\mathrm{d}x)^7$$

Hint. Use formula (3.1).

Lemma 3.5.4 *The maps J_7 and J_9 define non-trivial 2-cocycles on* Vect(S^1).

Proof Consider the (associative) product (3.26). The corresponding star-commutator (see (3.25)) is as follows:

$$[\phi, \psi]_* = \{\phi, \psi\} + \sum_{k=1}^{\infty} \left(\frac{t}{2}\right)^{2k+1} J_{2k+1}(\phi, \psi)$$

since the even-order transvectants are symmetric. Let us substitute vector fields X and Y to the above star-commutator.

Exercise 3.5.5 Check that the restrictions of the transvectants J_3 and J_5 to Vect$(S^1)^{\otimes 2}$ identically vanish.

It follows that the first non-zero terms in the series $[X, Y]_*$ are the usual commutator of vector fields $[X, Y]$ and the term proportional to $J_7(X, Y)$. The Jacobi identity for the star-commutator then implies that J_7 is, indeed, a 2-cocycle on Vect(S^1). For J_9, the Jacobi identity gives

$$J_9(X, [Y, Z]) + L_X^\lambda J_9(Y, Z) + J_3(X, J_7(Y, Z)) + \text{cycle}_{(X,Y,Z)} = 0.$$

Exercise 3.5.6 Check that

$$J_3(X, J_7(Y, Z)) + \text{cycle}_{(X,Y,Z)} = 0.$$

Thus J_9 is also a 2-cocycle.

These 2-cocycles are non-trivial; in other words, they cannot be written in the form (3.32). Indeed, for any linear map ℓ, the right hand side of (3.32) contains 0-jets of X and Y. Lemma 3.5.4 is proved. □

Theorem 3.5.2 now follows from the fact that transvectants are the unique sl(2, $\mathbb{R}$)-invariant bilinear maps from Vect(S^1) to $\mathcal{F}_\lambda(S^1)$; see Theorem 3.1.1. □

Comment

Extensions of Vect(S^1) by the modules $\mathcal{F}_\lambda(S^1)$ were classified in [223]; see also [72]. The 2-cocycles J_7 and J_9 were introduced in [169].

4
Vertices of projective curves

The four-vertex theorem is one of the first results of global differential geometry and global singularity theory. Already Apollonius had studied the caustic of an ellipse and observed its four cusps. Huygens also studied evolutes of plane curves and discovered their numerous geometric properties. Nevertheless the four-vertex theorem, asserting that the evolute of a plane oval has no fewer than four cusps, was discovered as late as 1909 by S. Mukhopadhyaya.

We start with a number of beautiful geometric results around the classic four-vertex theorem. We then prove a general theorem of projective differential geometry, due to M. Barner, and deduce from it various geometrical consequences, including the four-vertex and six-vertex theorems and the Ghys theorem on 4 zeroes of the Schwarzian derivative. The chapter concludes with applications and ramifications including discretizations, topological problems of wave propagation and connection with contact topology. A relates Section A.1 concerns the Sturm-Hurwitz-Kellogg theorem on the least number of zeroes of periodic functions.

This chapter illustrates the title of the book: it spans approximately 200 years, and we see how old and new results interlace with each other. The literature on the four-vertex theorem is immense. Choosing material for this chapter, we had to severely restrict ourselves, and many interesting results are not discuss here.

4.1 Classic four-vertex and six-vertex theorems

In this section we formulate the classic four-vertex and six-vertex theorems, their corollaries and refinements. The reader interested in their proofs should wait until Section 4.3.

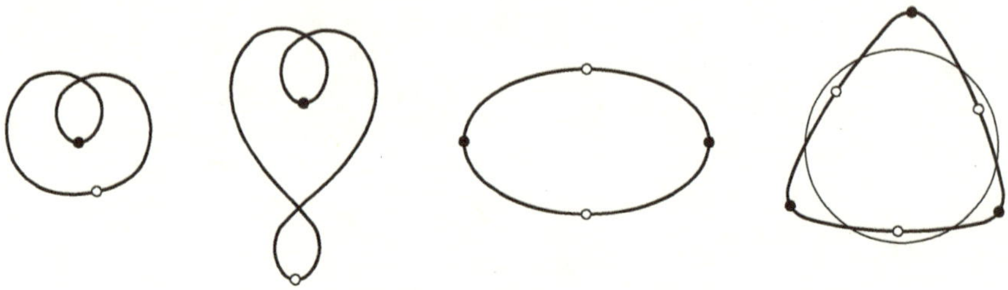

Fig. 4.1. Curves with 2, 4 and 6 vertices

The four-vertex theorem

Consider a smooth curve γ in the Euclidean plane. The osculating circle at a point $x \in \gamma$ is the circle tangent to the curve with order 2. One can say the osculating circle passes through three infinitesimally close points of the curve (i.e., has a "three-point contact" with the curve). The radius of the circle is reciprocal to the curvature of γ. The point x is called a *vertex* if the osculating circle approximates the curve abnormally well, that is, with order ≥ 3. Vertices are extrema of curvature.

The simplest version of the four-vertex theorem concerns smooth closed convex curves in the Euclidean plane with non-vanishing curvature, also called *ovals*.

Theorem 4.1.1 *An oval has at least four distinct vertices.*

The existence of two vertices is obvious: curvature has a maximum and a minimum. For self-intersecting curves, one can have only two vertices, see Figure 4.1.

One can strengthen the result as follows.

Theorem 4.1.2 *If an oval transversally intersects a circle in n points, then it has at least n distinct vertices.*

Of course, the number n in this theorem is even. Theorem 4.1.2 implies Theorem 4.1.1 since for any oval there is a circle intersecting it in at least four points.

Remark 4.1.3 Theorem 4.1.1 holds true for simple closed (but not necessarily convex) curves as well. We mention this stronger result in passing since its proof uses "ad hoc" arguments and is not directly related to the main ideas of this book.

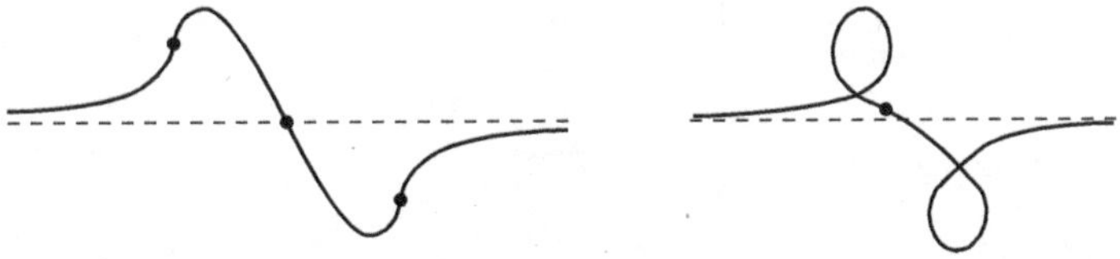

Fig. 4.2. Möbius theorem: example and counterexample

We mention another result in the same spirit, the Möbius theorem [149]: *a non-contractible closed simple curve in* $\mathbb{RP}^2$ *has at least three distinct inflection points*; see Figure 4.2. An inflection point is a point at which the curve has second-order contact with the tangent line.

Vertices and caustics

Given a smooth plane curve γ, consider the one-parameter family of its normal lines. The envelope of the normals is called the *caustic* or the evolute. This is a curve that can have singularities, generically, semicubical cusps. The caustic can also be defined as a locus of the curvature centers.

Vertices of a smooth plane curve correspond to cusps of its caustic, and the four-vertex theorem can be formulated as a four-cusps theorem for a caustic.

The caustics have been of great interest to physicists and mathematicians since the seventeenth century. Their numerous geometric properties are well known, and we will list here a few. The following statements used to be part of the calculus curriculum until the first half of the twentieth century.

Exercise 4.1.4 Show that:

(a) a caustic has no inflection points;
(b) one can reconstruct a curve from its caustic and an initial point;
(c) the number of cusps of the caustic of an oval is even and the alternating sum of the lengths of its smooth pieces equals zero;
(d) the osculating circles of an arc of a curve, free from vertices, are pairwise disjoint and nested.

Hint. (a) If a caustic had an inflection point, then there would be two perpendiculars to the curve at one of its points. (b) A curve is described by the free end of a non-stretchable string developing from the caustic. (c) This follows from this string construction. (d) This follows from the string construction and the triangle inequality.

We discuss symplectic and contact aspects of caustics in Section 4.6.

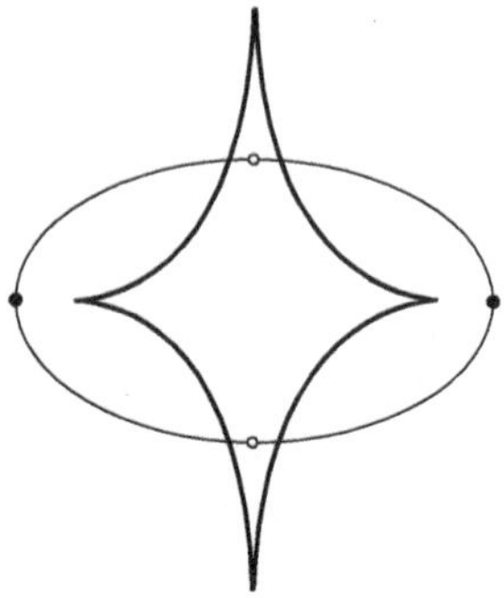

Fig. 4.3. Cusps of the caustic

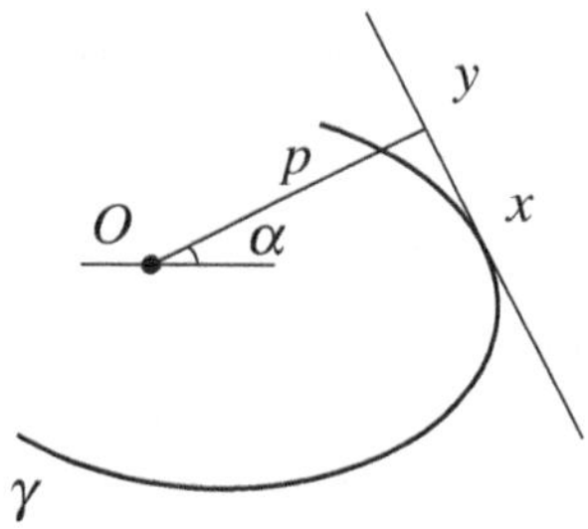

Fig. 4.4. Support function

Support function

It is convenient to define a convex curve by its *support function*. Let γ be a convex curve and x a point of γ; choose an origin O and draw the tangent line to γ at x. Drop the perpendicular Oy to the tangent line. The support function is the function $p(\alpha)$ where α is the direction of the vector Oy and p is its magnitude. If γ is (co)oriented, then p has a sign depending on whether the direction of the vector Oy is the same as the coorientation.

Exercise 4.1.5 Prove that translating the origin O through a vector (a, b) changes the support function by $a \cos \alpha + b \sin \alpha$.

Let us describe vertices in terms of the support function. We parameterize the curve γ by the angle $\alpha \in S^1$.

Lemma 4.1.6 *Vertices of γ correspond to the values of α for which*

$$p'''(\alpha) + p'(\alpha) = 0. \tag{4.1}$$

Proof Support functions of circles are $a\cos\alpha + b\sin\alpha + c$, where a, b and c are constants. Indeed, choosing O at the center of a circle, the support function is constant (the radius), and the general case follows from the above exercise.

Vertices are the points where the curve has a third-order contact with a circle. In terms of the support functions, it means that $p(\alpha)$ coincides with $a\cos\alpha + b\sin\alpha + c$ up to the third derivative. It remains to notice that linear harmonics $a\cos\alpha + b\sin\alpha + c$ satisfy (4.1) identically. □

Lemma 4.1.6 makes it possible to reformulate the four-vertex theorem as follows.

Theorem 4.1.7 *Let $p(\alpha)$ be a smooth 2π-periodic function. Then the equation $p'''(\alpha) + p'(\alpha) = 0$ has at least four distinct roots.*

See Appendix A.1 for a discussion of more general Sturm theorems of this kind. For a curious reader we mention more properties of support functions.

Exercise 4.1.8 Prove that:

(a) $|xy| = p'(\alpha)$;
(b) the radius of curvature at point x equals $p''(\alpha) + p(\alpha)$;
(c) the length of γ equals $\int p(\alpha)\mathrm{d}\alpha$;
(d) the area bounded by γ equals $\frac{1}{2}\int (p^2(\alpha) - (p'(\alpha))^2)\mathrm{d}\alpha$.

The six-vertex theorem

Let γ be a *convex* curve in $\mathbb{RP}^2$, that is, a smooth closed non-degenerate curve that intersects any projective line at most twice; cf. Section 2.3. An osculating conic at a point $x \in \gamma$ is a conic through x that has contact of order 4 (i.e., five-point contact) with γ.

Exercise 4.1.9 Show that five points in general position in $\mathbb{RP}^2$ determine a unique conic. Deduce the uniqueness of the osculating conic at every point.

A point is called *sextactic* if the osculating conic has contact of order ≥ 5 with γ at this point. Clearly, this is a projectively invariant notion.

Theorem 4.1.10 *A convex curve in $\mathbb{RP}^2$ has at least six distinct sextactic points.*

Similarly to the four-vertex theorem, one also has its strengthening.

Theorem 4.1.11 *If a convex curve in $\mathbb{RP}^2$ transversally intersects a conic at n points then it has at least n distinct sextactic points.*

As before, Theorem 4.1.10 follows from Theorem 4.1.11: it suffices to construct the conic through five points of γ, and it will intersect γ once again.

Sextactic points, cubic form and affine curvature

A sextactic point is a projective notion and one expects it to be related to projective differential invariants of curves. We have two such invariants, the projective curvature and the cubic form; see Section 1.4. As we already mentioned there, the cubic form of the curve vanishes in those points of the curve where the osculating conic is hyper-osculating, that is, in sextactic points.

Curiously, sextactic points are also affine vertices. As we proved in Theorem 2.3.1, a convex spherical curve lies in a hemisphere, and therefore a convex projective curve lies in an affine part of $\mathbb{RP}^2$. One then can consider its affine vertices.

Lemma 4.1.12 *Sextactic points are extremum points of the affine curvature.*

Proof The affine curvature was defined in Section 1.4 as a function of the affine parameter, so that the coordinates of the curve satisfied the equation $y'''(t) + k(t)y'(t) = 0$. According to formula (1.20), the cubic form is equal to $-(1/2)k'(t)(\mathrm{d}t)^3$. □

The following two statements are equivalent to the six-vertex theorem.

Corollary 4.1.13 *(i) The cubic form of a convex curve in $\mathbb{RP}^2$ has at least six distinct zeroes.*
(ii) The affine curvature of an oval in $\mathbb{R}^2$ has at least six distinct extrema.

Affine caustics

In affine geometry, as in Euclidean, one can define the caustic of a non-degenerate curve.

Let a parameterization $\gamma(t)$ satisfy $[\gamma'(t), \gamma''(t)] = 1$ for all t (an affine parameter, see Section 1.4). The lines $l(t)$ generated by the vectors $\gamma''(t)$ are called *affine normals* of the curve. The envelope of the affine normals is called the *affine caustic*.

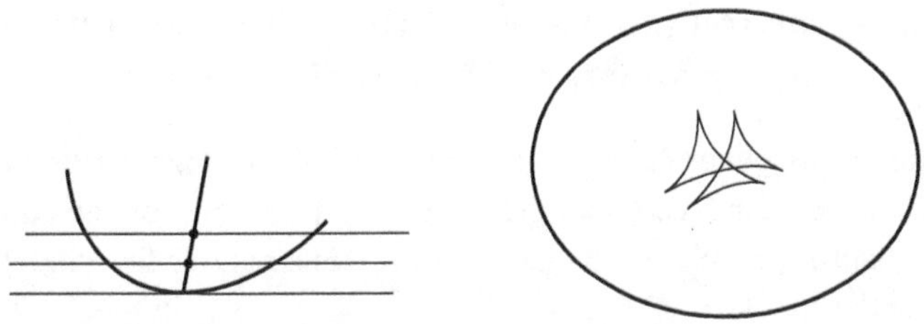

Fig. 4.5. Affine normal and affine caustic

Exercise 4.1.14 Show that the affine normal $l(t)$ is tangent to the curve that bisects the segments, bounded by the intersections of γ with the lines, parallel to the tangent line to γ at point $\gamma(t)$ see Figure 4.5.

Sextactic points of γ correspond to cusps of its caustic and the six-vertex theorem can be formulated as a 6-cusps theorem for the affine caustic.

Comment

The four-vertex theorem was proved by Mukhopadhyaya in [154]. In almost a hundred years since its publication this theorem has generated a thriving area of research connected, among other things, with contemporary symplectic topology and knot theory; see [11, 12].

Many interesting problems around the four-vertex theorem remain open; let us mention just one. In the posthumous "Lectures on Dynamics" by Jacobi one finds the following conjecture. Consider a generic smooth closed convex surface in $\mathbb{R}^3$, pick a generic point on it and consider the geodesic lines emanating from this point. The loci of the first, second, ... conjugate points are called first, second, ... caustics. Jacobi proved that each caustic has an even number of cusps; conjecturally, this number is not less than 4 for each caustic; see [12].

Sextactic points of algebraic curves in $\mathbb{CP}^2$ were thoroughly studied by Steiner, Plücker, Hesse and Cayley in the middle of the nineteenth century. Cayley proved that a generic curve of degree d has exactly $3d(4d-9)$ sextactic points. This resembles the better-known Plücker formula for the number of inflections, namely, $3d(d-2)$.

Sextactic points of smooth curves in $\mathbb{RP}^2$ were considered, for the first time, by Mukhopadhyaya [154]. Herglotz and Radon found a different proof of the six-vertex theorem, published in [26]; see also [88]. Recent results on the subject were obtained by Thorbergsson and Umehara in [217]. In particular, they show that there exist convex degenerate curves with just two sextactic points.

4.2 Ghys' theorem on zeroes of the Schwarzian derivative and geometry of Lorentzian curves

Among numerous generalizations and analogs of the four-vertex theorem, a theorem discovered by E. Ghys stands out. This is one of the most recent and most beautiful results in the area, and it concerns our favorite object, the Schwarzian derivative.

Four zeroes of the Schwarzian

Consider a diffeomorphism $f : \mathbb{RP}^1 \to \mathbb{RP}^1$. The Ghys theorem asserts the following.

Theorem 4.2.1 *The Schwarzian derivative $S(f)$ vanishes in at least four distinct points.*

As in Section 4.1, one also has its strengthening.

Theorem 4.2.2 *If the diffeomorphism f has n non-degenerate fixed points, then $S(f)$ vanishes in at least n distinct points.*

As before, the number n is even. A fixed point is non-degenerate if $f' \neq 1$ at this point. Theorem 4.2.2 implies Theorem 4.2.1 since one can find a projective transformation g such that $g \circ f$ has three (and therefore four) fixed points. One then concludes by formula (1.9).

The Schwarzian derivative of f was defined in Section 1.3 to measure the failure of f to preserve the cross-ratio. On the other hand, the diffeomorphisms preserving the cross-ratio are precisely the elements of $\mathrm{PGL}(2, \mathbb{R})$. At any point x of $\mathbb{RP}^1$, there exists a unique projective transformation $g \in \mathrm{PGL}(2, \mathbb{R})$ that approximates f in x up to order 2 (has three-point contact with f). Let us call this projective transformation *osculating* f at x. The points in which the Schwarzian $S(f)$ vanishes are precisely the points in which this approximation is at least of order 3.

Vertices in Lorentz geometry

Theorem 4.2.1 is clearly an analog of the four-vertex theorem. In order to better understand this viewpoint, let us assign a curve to a diffeomorphism of $\mathbb{RP}^1$. The choice is natural: the graph of the diffeomorphism is a curve in $\mathbb{RP}^1 \times \mathbb{RP}^1$. Furthermore, we would like to define a geometry on $\mathbb{RP}^1 \times \mathbb{RP}^1$ in which the graph of a projective transformation is a circle.

Consider the flat Lorentz metric $\mathrm{g} = dxdy$ on $\mathbb{RP}^1 \times \mathbb{RP}^1$, where x and y are affine coordinates on $\mathbb{RP}^1 = \mathbb{R} \cup \{\infty\}$.

Exercise 4.2.3 Show that the circles in the flat Lorentz plane are the hyperbolas $(x-a)(y-b) = c$, where a, b, c are constants and $c \neq 0$.

Note that the hyperbola $(x-a)(y-b) = c$ is the graph of the linear-fractional transformation

$$y = \frac{bx + c - ab}{x - a}.$$

Thus, the flat Lorentz geometry is what we need.

The graph of an orientation-preserving diffeomorphism is a *wordline* (or a space-like curve).

Lemma 4.2.4 *Let t be the Lorentz arc-length*[1] *of a wordline and $\kappa(t)$ its Lorentz curvature. Then one has*

$$\mathrm{d}\kappa\,\mathrm{d}t = \frac{1}{2}\,S(f). \tag{4.2}$$

Proof Let J be the linear operator $(x, y) \mapsto (-x, y)$. Then $J(v)$ is orthogonal to v, and $\mathrm{g}(J(u), J(v)) = -\mathrm{g}(u, v)$ for all vectors u and v. This is the Lorentz right-angle rotation.

Let $\gamma(t)$ be the curve $y = f(x)$. Denote $\mathrm{d}/\mathrm{d}t$ by dot and $\mathrm{d}/\mathrm{d}x$ by prime. One easily finds that

$$\dot{\gamma}(t) = \frac{1}{f'^{1/2}}\,(1, f') \quad \text{and} \quad \ddot{\gamma}(t) = \frac{1}{2}\,\frac{f''}{f'^{3/2}}\,J(\dot{\gamma}(t)).$$

Similarly to the familiar Euclidean case, the above coefficient

$$\kappa = \frac{1}{2}\,\frac{f''}{f'^{3/2}} \tag{4.3}$$

is, by definition, the Lorentz curvature of γ. Recall that

$$S(f) = \left(\frac{f'''}{f'} - \frac{3}{2}\left(\frac{f''}{f'}\right)^2\right)(\mathrm{d}x)^2.$$

Differentiating (4.3), one obtains (4.2). □

[1] Also called *proper time*.

It follows that zeroes of the Schwarzian derivative are Lorentz vertices and the Ghys theorem is the four-vertex theorem in Lorentz geometry.

Exercise 4.2.5 Check that equation (4.2) holds for the Lorentz metrics

$$g = \frac{dxdy}{(axy + bx + cy + d)^2} \tag{4.4}$$

where a, b, c and d are arbitrary real constants.

This is a metric of constant scalar curvature $R = 8(ad - bc)$ defined on $\mathbb{RP}^1 \times \mathbb{RP}^1 - \mathbb{RP}^1$, the complement of the graph of the linear-fractional transformation $y = -(bx + d)/(ax + c)$; see [52].

The four-vertex theorem in the hyperbolic plane

The classic four-vertex theorem holds for Riemannian metrics of constant curvature (on the sphere and in the hyperbolic plane). In fact, one can deduce the four-vertex theorem in the hyperbolic plane from the Ghys theorem.

Consider the Klein–Beltrami (or projective) model of hyperbolic geometry. The hyperbolic plane H^2 is represented by the interior of the unit circle S^1 ("circle at infinity"), straight lines – by chords of this circle, and the distance between points x and y is given by the formula

$$d(x, y) = \frac{1}{2} \ln[a, x, y, b], \tag{4.5}$$

where a and b are the intersection points of the line xy with S^1 and $[a, x, y, b]$ is the cross-ratio; see formula (1.5). Isometries of H^2 are the projective transformations of the plane preserving the boundary circle.

The circle at infinity S^1 has a natural projective structure of a conic in the projective plane (discussed in Exercise 1.4.2). Isometries of H^2 preserve this projective structure.

Let γ be a convex oriented curve in H^2. We define a diffeomorphism of the circle at infinity as follows. Given $x \in S^1$, consider the oriented tangent line to γ and let $f(x)$ be the second intersection point of this line with S^1; see Figure 4.6. This defines a diffeomorphism f of S^1.

Proposition 4.2.6 *Vertices of the curve γ correspond to zeroes of the Schwarzian $S(f)$.*

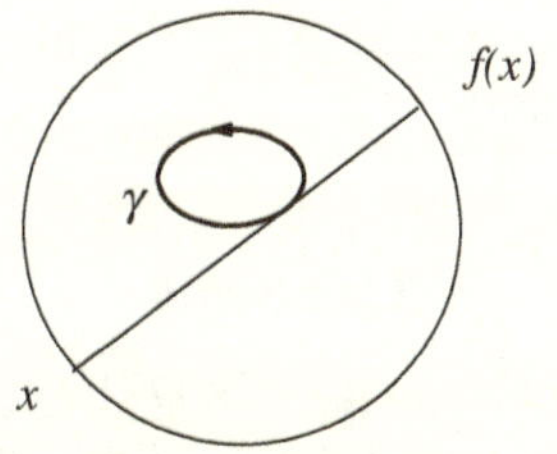

Fig. 4.6. Circle diffeomorphism from a curve

Proof A vertex is a four-point tangency of γ with a curve of constant curvature κ. Such a curve is a circle if $\kappa > 1$, a horocycle if $\kappa = 1$ and an equidistant curve if $\kappa < 1$. For a reference on hyperbolic geometry, see, for example, [21].

A zero of the Schwarzian $S(f)$ occurs when f is abnormally well approximated by a projective transformation of the circle at infinity, which is a restriction of a projective transformation of the plane, preserving this circle. Therefore, it suffices to show that, if f is a projective transformation, that is, an isometry of the hyperbolic plane, then γ is a curve of constant curvature.

Recall the classification of isometries: an elliptic isometry has a unique fixed point and is a rotation of the hyperbolic plane, a parabolic isometry has a unique fixed point at infinity, and a hyperbolic isometry has two fixed points at infinity.

Consider the case when f is an elliptic isometry. Without loss of generality, it is a rotation about the center of the unit disc. Then γ is a circle, concentric with the unit disc, and this settles the case $\kappa > 1$.

Now let f be a parabolic isometry. Let us consider the upper half-plane model in which the isometries are represented by elements of $SL(2, \mathbb{R})$, the geodesics by half-circles perpendicular to the absolute, the x-axis, and the horocycles by horizontal lines. Without loss of generality, f is a horizontal parallel translation $(x, y) \mapsto (x + c, y)$. The family of geodesics, connecting points of the absolute with their images, consists of Euclidean congruent half-circles, translated along the x-axis. The envelope of these geodesics is a horizontal line, that is, a horocycle. This settles the case $\kappa = 1$.

Finally, let f be a hyperbolic isometry. Working again in the upper half-plane model, we may assume that f is a dilation $(x, y) \mapsto (cx, cy)$. The family of geodesics, connecting points of the absolute with their images, consists of Euclidean homothetic half-circles, whose envelope is the line $y = Cx$. This line is an equidistant curve with constant curvature $\kappa < 1$. This completes the proof. □

The Ghys theorem provides four zeroes of a diffeomorphism of $\mathbb{RP}^1$. According to Exercise 1.4.2, the projective structure on the circle at infinity coincides with the standard projective structure of $\mathbb{RP}^1$. This gives four zeroes of $S(f)$ and therefore four vertices of γ.

Comment

E. Ghys presented his theorem in the talk [79]; see also [171]. The result of Exercise 4.2.5 is borrowed from [56]; condition (4.4) is also necessary for equation (4.2) to hold. The four-vertex theorem in Lorentz geometry was found by A. Kneser [120].

Proposition 4.2.6 and the deduction of the four-vertex theorem from the Ghys theorem is due to D. Singer [192]; see also [211] for a relation between circle diffeomorphisms and curves in the hyperbolic plane.

The group of conformal transformations of the metric (4.4) is isomorphic to $\mathrm{Diff}_+(\mathbb{RP}^1)$; see [91]. In [53], the space of metrics, conformally equivalent to (4.4), is related to coadjoint orbits of the Virasoro algebra.

4.3 Barner's theorem on inflections of projective curves

In this section we formulate and prove a general theorem on curves in $\mathbb{RP}^n$. This result will be our powerful tool throughout this chapter. The section starts with the definition of strictly convex projective curves after M. Barner. We then give a detailed proof of the Barner theorem.

Strictly convex curves after Barner

Unlike the preceding sections, the curves considered here are not necessarily non-degenerate.

A curve $\gamma \subset \mathbb{RP}^n$ is called *strictly convex* if, for every $(n-1)$-tuple of points of γ, there exists a hyperplane through these points that does not intersect γ anywhere else. In this definition, $(n-1)$-tuple may contain the same point $k > 1$ times. In this case, the curve should have contact of order $k-1$ with the respective hyperplane.

Remark 4.3.1 The term "strictly convex" should not be confused with the term "convex" widely used in this book (see Section 2.3). We will discuss the relation between convex curves and strictly convex in Section 4.4.

Let us list a couple of simple geometric properties of strictly convex curves.

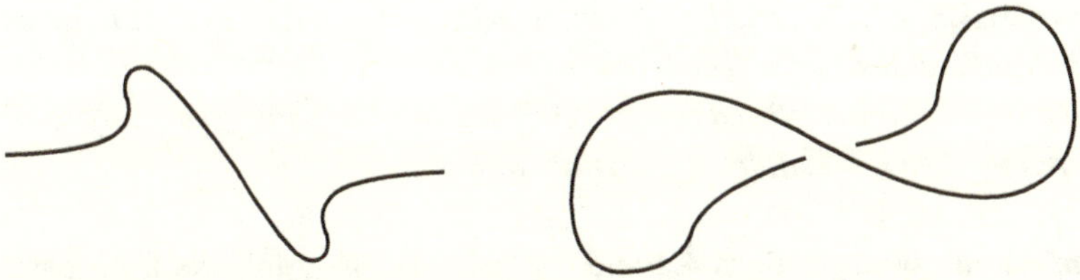

Fig. 4.7. Strictly convex curves in $\mathbb{RP}^2$ and $\mathbb{RP}^3$

Exercise 4.3.2 Show that:

(a) the osculating flag $F_1 \subset \cdots \subset F_{n-1}$ (i.e., the first $n-1$ spaces in the flag (1.1)) of a strictly convex curve is full;
(b) a strictly convex curve is contractible if n is odd and non-contractible if n is even.

Property (a) guarantees that the osculating hyperplane is well defined at every point of γ.

The notion of strict convexity is quite remarkable. Let us prove here a useful property of strictly convex curves. Let $C(x)$ be the osculating codimension 2 space of γ at a point x. There exists a hyperplane, $H(x)$, containing $C(x)$ and intersecting γ nowhere else. Let us call $H(x)$ a *Barner hyperplane*.

Lemma 4.3.3 *The set of Barner hyperplanes at x is contractible for every $x \in \gamma$.*

Proof The space of all hyperplanes containing $C(x)$ is a circle. Let H_1 and H_2 be two Barner hyperplanes. Together, they separate $\mathbb{RP}^n$ into two components. Since the curve $\gamma \setminus \{x\}$ is connected, it belongs to one of the components. We can rotate H_1 to H_2 in the other component. Therefore, the space of Barner hyperplanes at x is a proper connected subset of a circle, that is, an interval. □

Barner's theorem

We are ready to formulate and prove the Barner theorem, one of the strongest results on global geometry of curves.

An *inflection* (or *flattening*) point of a space curve is a point in which the curve fails to be non-degenerate. In other words, the osculating flag (1.1) is not full. By Exercise 4.3.2 part (a), inflection points of a strictly convex curve are stationary points of the osculating hyperplane.

Theorem 4.3.4 *(i) A strictly convex closed curve in $\mathbb{RP}^n$ has at least $n+1$ distinct inflections.*
(ii) If a strictly convex closed curve in $\mathbb{RP}^n$ transversally intersects a hyperplane in k points then it has at least k distinct inflections.

Proof Similarly to Sections 4.1 and 4.2, part (i) easily follows from part (ii). Pick n generic points on the curve γ and consider the hyperplane through these points. Exercise 4.3.2 part (b) implies that γ has to intersect the hyperplane in at least $n+1$ points. Indeed, the number of intersections is even if n is odd, and odd if n is even.

To prove (ii), let us use induction on n. The case of $n = 1$ is the following statement: if γ is a curve in $\mathbb{RP}^1$ that does not pass through some point and passes through another point k times then γ has k singularities. This follows from the Rolle theorem.

Consider the induction step from $n-1$ to n. Let x be one of the intersection points with the hyperplane, and let π be the projection $\mathbb{RP}^n \to \mathbb{RP}^{n-1}$ from x. We may assume that all intersection points are not inflection points of γ, otherwise one may slightly perturb the hyperplane. Denote the curve $\pi(\gamma)$ by $\bar{\gamma}$.

Lemma 4.3.5 *$\bar{\gamma}$ is a smooth strictly convex closed curve.*

Proof The points at which the projection of γ fails to be smooth are those points $p \in \gamma$ at which the tangent line to γ passes through x. The existence of such points contradicts strict convexity. Indeed, choose arbitrary points $y_1, \dots, y_{n-3}$ on γ and consider the $(n-1)$-tuple $(x, p, y_1, \dots, y_{n-3})$. Any hyperplane through these points contains the tangent line px and thus intersects γ with total multiplicity $\geq n$.

It remains to investigate the image of γ in a neighborhood of x. We define the projection $\pi(x)$ as the tangent line to γ at x. This makes $\bar{\gamma}$ closed. Moreover, x is not an inflection point. Introduce a local parameter t near x. The curve $\gamma(t)$ at x is given by

$$\gamma(t) = (t + O(t^2), t^2 + O(t^3), \dots, t^n + O(t^{n+1}))$$

in some affine coordinate system. Hence the curve $\bar{\gamma}$ at $\pi(x)$ is given by

$$\bar{\gamma}(t) = (t + O(t^2), t^2 + O(t^3), \dots, t^{n-1} + O(t^n)).$$

We have proved that $\bar{\gamma}$ is smooth.

Given $y_1, \dots, y_{n-2} \in \bar{\gamma}$, consider their preimages $x_i = \pi^{-1}(y_i)$. There exists a hyperplane in $\mathbb{RP}^n$ through points $x, x_1, \dots, x_{n-2}$ that does not intersect γ anywhere else, and its projection to $\mathbb{RP}^{n-1}$ is the desired hyperplane therein. $\square$

By the induction assumption, since $\bar{\gamma}$ intersects a hyperplane $(k-1)$ times, it has $(k-1)$ inflections.

Lemma 4.3.6 *An inflection of $\bar{\gamma}$ corresponds to an osculating hyperplane of γ passing through x.*

Proof Lift the curves γ and $\bar{\gamma}$ to $\mathbb{R}^{n+1}$ and $\mathbb{R}^n$, respectively, as Γ and $\bar{\Gamma}$. Choose a parameter t on our curves. Let X be the lift of point x. The vectors $\bar{\Gamma}, \bar{\Gamma}', \bar{\Gamma}'', \ldots, \bar{\Gamma}^{(n-1)}$ are the projections of $\Gamma, \Gamma', \Gamma'', \ldots, \Gamma^{(n-1)}$. By Exercise 4.3.2 part (a), the latter vectors are linearly independent. These vectors span a hyperplane in $\mathbb{R}^{n+1}$ whose projection to $\mathbb{R}^n$ is not bijective only if it contains X. $\square$

Let $\pi(y)$ be an inflection point of $\bar{\gamma}$. We want to investigate the curve γ in a neighborhood of point y. Let t be a local parameter with $\gamma(0) = y$ and let $Z(t) = X - \Gamma(t)$, where $X \in \Gamma$ is the lift of x. If y is not an inflection point of γ then the vectors $\Gamma(t), \ldots, \Gamma^{(n)}(t)$ are linearly independent, and one can write

$$Z(t) = \alpha_0(t)\,\Gamma(t) + \cdots + \alpha_n(t)\,\Gamma^{(n)}(t). \tag{4.6}$$

Since $t = 0$ corresponds to inflection of $\bar{\gamma}$, Lemma 4.3.6 ensures that $\alpha_n(0) = 0$, and one has:

$$Z(0) = \alpha_0(0)\,\Gamma(0) + \cdots + \alpha_{n-1}(0)\,\Gamma^{(n-1)}(0). \tag{4.7}$$

We are interested in the orientations of two frames: $(\Gamma(t), \ldots, \Gamma^{(n)}(t))$ and $(\Gamma(t), \ldots, \Gamma^{(n-1)}(t), Z(t))$.

Lemma 4.3.7 *In a small neighborhood of y, one has*

$$\left|\Gamma(t)\ldots\Gamma^{(n-1)}(t)\,Z(t)\right| = -t\,\alpha_{n-1}(0)\left|\Gamma(0)\ldots\Gamma^{(n)}(0)\right| + O(t^2). \tag{4.8}$$

Proof Let us write $Z(t) = Z(0) - (\Gamma(t) - \Gamma(0))$ and substitute in the left hand side of (4.8). Then one has

$$\left|\Gamma(t)\ldots\Gamma^{(n-1)}(t)\,Z(t)\right| = \left|\Gamma(t)\ldots\Gamma^{(n-1)}(t)\,Z(0)\right| + O(t^2)$$

since $\Gamma(t) - \Gamma(0) = \Gamma'(0)t + O(t^2)$.

Furthermore, substitute $\Gamma^{(i)}(t) = \Gamma^{(i)}(0) + \Gamma^{(i+1)}(0)t + O(t^2)$ and expression (4.7) on the right hand side of the last formula and collect terms. The result follows. □

The point $\pi(x)$ is not an inflection point of $\bar{\gamma}$; see the proof of Lemma 4.3.5. Let $0 \le t \le 1$ be a parameter on γ, such that $x = \gamma(0) = \gamma(1)$ and $0 < t_1 < \ldots < t_{k-1} < 1$ are the values of the parameter, corresponding to the inflection points of $\bar{\gamma}$.

Lemma 4.3.8 *Each interval* (t_i, t_{i+1}) *contains an inflection of* γ.

Proof Note first that the left hand side determinant in formula (4.8) does not change sign on the interval since the curve $\bar{\gamma}$ does not have inflections on this interval; cf. Lemma 4.3.6.

Next, we claim that the constants $\alpha_{n-1}(t_i)$ and $\alpha_{n-1}(t_{i+1})$ do not vanish. Indeed, if $\alpha_{n-1}(0) = 0$ in (4.7), then every Barner hyperplane at y that contains $\gamma'(0), \ldots, \gamma^{(n-2)}(0)$ passes through an additional point $x \in \gamma$. This contradicts strict convexity.

Let us prove that $\alpha_{n-1}(t_i)$ and $\alpha_{n-1}(t_{i+1})$ have the same sign. For every $t \in (t_i, t_{i+1})$, choose a Barner hyperplane through $\gamma(t)$ that contains $\gamma'(t), \ldots, \gamma^{(n-2)}(t)$ and lift it to $\mathbb{R}^{n+1}$ as $H(t)$. By Lemma 4.3.3, we may assume that $H(t)$ depends continuously on t.

Assume that $\alpha_{n-1}(t_i)$ and $\alpha_{n-1}(t_{i+1})$ have opposite signs. Then the vectors $Z(t)$ and $\Gamma^{(n-1)}(t)$ lie on one side of $H(t)$ for one of the boundary values $\{t_i, t_{i+1}\}$ and on the opposite sides of $H(t)$ for the other. Therefore, either $Z(t)$ or $\Gamma^{(n-1)}(t)$ belong to $H(t)$ at some $t \in (t_i, t_{i+1})$. This contradicts strict convexity.

The parameter t in formula (4.8) is a local parameter in a vicinity of an inflection point of $\bar{\gamma}$. Applying this formula twice, on small intervals $(t_i, t_i + \varepsilon)$ and $(t_i - \varepsilon, t_i)$, we conclude that the determinant $|\Gamma(t) \ldots \Gamma^{(n)}(t)|$ has opposite signs at the boundary values t_i and t_{i+1}, that is, vanishes at some intermediate point. □

To complete the proof of the Barner theorem, it remains to consider two extreme intervals.

Lemma 4.3.9 *Each of the intervals* $(0, t_1)$ *and* $(t_{k-1}, 1)$ *contains an inflection of* γ.

Proof Consider the interval $(0, t_1)$; the situation with the other interval, $(t_{k-1}, 1)$ is exactly the same. The proof proceeds along the same lines as that of the preceding lemma.

Formula (4.8) is valid in a neighborhood of t_1 but not near 0. Instead, one has:

$$\left|\Gamma(t)\ldots\Gamma^{(n-1)}(t)\,Z(t)\right| = (-1)^n\,\frac{t^n}{n!}\,\left|\Gamma(0)\ldots\Gamma^{(n)}(0)\right| + O(t^{n+1}) \qquad (4.9)$$

as follows from the Taylor formula. Let us write an explicit expression (4.6) for $t \geq 0$ sufficiently close to 0. The vector $Z(t)$ is defined as $\Gamma(0) - \Gamma(t)$. Using the Taylor expansion

$$\Gamma(0) = \sum_{i=0}^{n}(-1)^i\,\frac{t^i}{i!}\,\Gamma^{(i)}(t) + O(t^{n+1}),$$

it follows that

$$\alpha_{n-1}(t) = (-1)^{n-1}\,\frac{t^{n-1}}{(n-1)!} + O(t^{n+1}).$$

Hence, the sign of $\alpha_{n-1}(t)$ at a small interval $(0, \varepsilon)$ is $(-1)^{n-1}$.

The arguments in the proof of Lemma 4.3.8 still apply and therefore $\alpha_{n-1}(t)$ also has the sign $(-1)^{n-1}$ left of t_1. The sign of the determinant $\left|\Gamma(t)\ldots\Gamma^{(n-1)}(t)\,Z(t)\right|$ remains the same on the whole interval $(0, t_1)$.

Finally, by formulæ (4.8) and (4.9), we conclude that the determinant $\left|\Gamma(t)\ldots\Gamma^{(n)}(t)\right|$ changes sign on the interval $(0, t_1)$. □

This completes the induction step, and Theorem 4.3.4 follows. □

Comment

Theorem 4.3.4 was published in [18], and this paper is not sufficiently well known. We believe that Barner's theorem deserves more attention. Our proof follows Barner's idea but is more detailed.

4.4 Applications of strictly convex curves

In this section we deduce Theorems 4.1.2, 4.1.11 and 4.2.2 from the Barner theorem. Using the Veronese map, we reformulate theorems on vertices of a plane curve in terms of inflections of the corresponding curve in $\mathbb{RP}^n$. The results then follow from the fact that the image of a convex curve is strictly convex. Another significant application of the Barner theorem is related to the notion of convex curves in $\mathbb{RP}^n$.

Deducing theorems on vertices

The scheme of the proof of Theorems 4.1.2, 4.1.11 and 4.2.2 is the same. We will consider one of them, the theorem on sextactic points, in detail.

Consider the *Veronese map* $\mathcal{V} : \mathbb{RP}^2 \to \mathbb{RP}^5$ of degree 2:

$$\mathcal{V} : (x : y : z) \mapsto (x^2 : y^2 : z^2 : xy : yz : zx). \tag{4.10}$$

Let γ be a convex curve in $\mathbb{RP}^2$ (recall that "convex" means non-degenerate and intersecting every line in at most two points; see Section 2.3).

Lemma 4.4.1 *The curve $\mathcal{V}(\gamma)$ is strictly convex.*

Proof The Veronese map establishes a one-to-one correspondence between conics in $\mathbb{RP}^2$ and hyperplanes in $\mathbb{RP}^5$: the image of the conic is the intersection of a hyperplane with the quadratic surface $\mathcal{V}(\mathbb{RP}^2)$. Given four points on $\mathcal{V}(\gamma)$, consider their preimages, four points on γ. Two straight lines in $\mathbb{RP}^2$ through these points have no other intersections with γ. A pair of lines is a (degenerate) conic, and the corresponding hyperplane in $\mathbb{RP}^5$ intersects $\mathcal{V}(\gamma)$ in exactly four given points. □

In order to apply Barner's theorem we need to interpret sextactic points of γ in terms of $\mathcal{V}(\gamma)$.

Lemma 4.4.2 *Sextactic points of γ correspond to inflection points of $\mathcal{V}(\gamma)$.*

Proof A sextactic point of γ is a point of fifth-order contact with a conic. The corresponding point of $\mathcal{V}(\gamma)$ is a point of fifth-order contact with the respective hyperplane in $\mathbb{RP}^5$, that is, an inflection point. □

Combining the two lemmas, one deduces Theorem 4.1.11 from Barner's theorem.

To prove Theorems 4.1.2 and 4.2.2, one considers the Veronese map from $\mathbb{RP}^2$ to $\mathbb{RP}^3$:

$$(x : y : z) \mapsto (x^2 + y^2 : z^2 : yz : zx) \tag{4.11}$$

and the Segre map from $\mathbb{RP}^1 \times \mathbb{RP}^1$ to $\mathbb{RP}^3$

$$((x_1 : y_1), (x_2 : y_2)) \mapsto (x_1x_2 : x_1y_2 : y_1x_2 : y_1y_2), \tag{4.12}$$

respectively.

Exercise 4.4.3 Formulate and prove analogs of Lemmas 4.4.1 and 4.4.2 for the maps (4.11) and (4.12).

Strict convexity and convexity

A curve γ in $\mathbb{RP}^n$ is called *convex* if every hyperplane intersects γ in at most n points, multiplicities counted. We already considered convex curves in $\mathbb{RP}^2$.in Sections 2.3 and 4.1.

Exercise 4.4.4 Show that:

(a) a convex curve is non-degenerate;
(b) a convex curve in $\mathbb{RP}^n$ is contractible for n even and non-contractible for n odd.

Example 4.4.5 The curves

$$(1 : \cos t : \sin t : \cos 2t : \sin 2t : \cdots : \sin nt) \subset \mathbb{RP}^{2n},$$

where $0 \leq t < 2\pi$, and

$$(\cos t : \sin t : \cos 3t : \sin 3t : \cdots : \sin (2n-1)t) \subset \mathbb{RP}^{2n-1},$$

where $0 \leq t < \pi$, are convex.

The reader will check that this curve is nothing else but the normal curve (2.11) with the parameters related by $x = \tan t/2$, in the first case, and $x = \tan t$, in the second.

The relation between convexity and strict convexity is as follows.

Lemma 4.4.6 *If the projection of a curve γ in $\mathbb{RP}^n$ to $\mathbb{RP}^{n-1}$ from a point O outside of γ is convex then γ is strictly convex.*

Proof Consider $n-1$ points on γ. Then the hyperplane through these points and O does not intersect γ anywhere else. Otherwise, the projection of this hyperplane would intersect the projection of γ in more than $n-1$ points. □

The Barner theorem now implies the following statement.

Corollary 4.4.7 *If a curve in $\mathbb{RP}^n$ has a convex projection to a hyperplane then it has at least $n+1$ inflection points.*

Convex curves and differential operators

Non-degenerate projective curves and linear differential operators are closely related; see Section 2.2. Convex curves correspond to a special class of differential operators.

A differential operator

$$A = \frac{\mathrm{d}^{n+1}}{\mathrm{d}x^{n+1}} + a_n(x)\frac{\mathrm{d}^n}{\mathrm{d}x^n} + \cdots + a_1(x)\frac{\mathrm{d}}{\mathrm{d}x} + a_0(x)$$

with 2π-periodic coefficients is called *disconjugate* if: (1) every solution of the equation $Af = 0$ has at most n zeroes on $[0, 2\pi)$, multiplicities counted; (2) every solution satisfies the condition $f(x+2\pi) = (-1)^n f(x)$.

Exercise 4.4.8 Show that the correspondence of Section 2.2 between closed curves in $\mathbb{RP}^n$ and differential operators of the form (2.7) associates disconjugate operators with convex curves.

For instance, the first curve from Example 4.4.5 corresponds to the differential operator

$$A_n = \frac{\mathrm{d}}{\mathrm{d}t}\left(\frac{\mathrm{d}^2}{\mathrm{d}t^2} + 1\right)\left(\frac{\mathrm{d}^2}{\mathrm{d}t^2} + 4\right)\cdots\left(\frac{\mathrm{d}^2}{\mathrm{d}t^2} + n^2\right) \tag{4.13}$$

which is just the Bol operator $\mathrm{D}_{2n+1} = \mathrm{d}^{2n+1}/\mathrm{d}x^{2n+1}$ from Theorem 2.1.2 where $x = \tan t/2$.

Theorem 4.4.9 *If a differential operator A of order $n+1$ is disconjugate and f is a smooth function such that $f(x+2\pi) = (-1)^n f(x)$ then the function Af has at least $n+2$ distinct zeroes.*

Proof Consider the differential equation $A\phi = 0$ and choose a basis of solutions $\phi_1, \ldots, \phi_{n+1}$. The curve $\gamma = (\phi_1 : \cdots : \phi_{n+1}) \subset \mathbb{RP}^n$ is convex since A is disconjugate. The curve $\widetilde{\gamma} = (\phi_1 : \cdots : \phi_{n+1} : f) \subset \mathbb{RP}^{n+1}$ is strictly convex by Lemma 4.4.6.

We claim that inflection points of γ correspond to zeroes of the function Af. Indeed, inflection points of γ are the points at which the Wronski determinant

$$W(\widetilde{\gamma}) = \begin{vmatrix} \phi_1 & \cdots\ \phi_{n+1} & f \\ & \cdots & \\ \phi_1^{(n+1)} & \cdots\ \phi_{n+1}^{(n+1)} & f^{(n+1)} \end{vmatrix}$$

vanishes. Using the equation $A\phi_i = 0$ and adding i-th row, multiplied by a_i, to the last row, one obtains

$$W(\widetilde{\gamma}) = W(\gamma)\,Af.$$

It remains to notice that $W(\gamma) \neq 0$ since γ is a non-degenerate curve.

The result follows now from the Barner theorem. □

We encountered a particular case of Theorem 4.4.9 in Theorem 4.1.7 where the differential operator was A_1; see (4.13).

We cannot help mentioning the following classic result.

Theorem 4.4.10 *If γ is a convex curve in $\mathbb{RP}^n$, then the projectively dual curve γ^* is also convex.*

The proof is based on the fact that an operator on an interval is disconjugate if and only if it can be decomposed into a product of first-order differential operators (we do not give here a proof which involves a substantial amount of analysis). Then the dual operator is also decomposable. The dual operator corresponds to the dual curve (see Theorem 2.2.6), which is therefore convex.

Extactic points of plane curves

We already studied approximations of curves in $\mathbb{RP}^2$ by conics and by straight lines; cf. Theorem 4.1.10 on six sextactic points and the Möbius theorem on inflection points. In these cases we were concerned with abnormally fine approximations of the curve by algebraic curves of degree 2 and 1, respectively.

Consider the space of algebraic curves of degree n in $\mathbb{RP}^2$. It has the dimension $d(n) = n(n+3)/2$. A plane curve γ has a unique osculating curve of degree n at every point. This osculating algebraic curve has a $d(n)$-point contact with γ at a generic point. A point of γ is called *n-extactic* if the contact is of order higher than $d(n)$.

It is an intriguing question to estimate below the number of n-extactic points of plane curves (or of some special classes of plane curves).

Theorem 4.4.11 *A plane curve sufficiently close to the oval of an irreducible cubic has at least ten 3-extactic points.*

Proof Consider the Veronese map $\mathcal{V} : \mathbb{RP}^2 \to \mathbb{RP}^9$

$$\mathcal{V} : (x : y : z) \mapsto (x^3 : x^2y : x^2z : xy^2 : xyz : xz^2 : y^3 : y^2z : yz^2 : z^3).$$

Let γ_0 be the oval of a cubic and γ is its small perturbation. Then $\mathcal{V}(\gamma_0)$ lies in a hyperplane $\mathbb{RP}^8$.

We claim that $\mathcal{V}(\gamma_0)$ is convex in this hyperplane. Indeed, the intersection of $\mathcal{V}(\gamma_0)$ with a hyperplane corresponds to the intersection of γ_0 with a cubic. By the Bezout theorem, this intersection consists of at most nine points. But the number of intersections with an oval is even. Therefore, it is at most 8.

The projection of $\mathcal{V}(\gamma) \subset \mathbb{RP}^9$ to $\mathbb{RP}^8$ is a small perturbation of γ_0. Therefore this projection is convex and so $\mathcal{V}(\gamma)$ is strictly convex; see Lemma 4.4.6. It remains to use the Barner theorem and the fact that inflection points of $\mathcal{V}(\gamma)$ correspond to 3-extactic points of γ. □

Comment

The unified proof of Theorems 4.1.2 and 4.1.11 as consequences of Theorem 4.3.4 is due to Barner [18]. A similar proof of Theorem 4.2.2 was given in [204].

The notion of convex curves in $\mathbb{RP}^n$ is classical. Coordinates of a lift of a convex curve to $\mathbb{R}^{n+1}$ form a Chebyshev systems of functions on S^1. Chebyshev systems have been thoroughly studied; see, for example, [46, 104] for a detailed proof of Theorem 4.4.10 and many other results. Theorem 4.4.11 and the notion of "extactic" points are due to V. Arnold [14]; the term was suggested by D. Eisenbud. Our proof of Theorem 4.4.9 follows [204].

The motivation for the study of strictly convex curves was their applications to the classic theorems on vertices. We have seen that this notion has a wealth of other applications and deserves further study.

4.5 Discretization: geometry of polygons, back to configurations

In this section we discuss a discrete version of projective differential geometry of curves. We consider closed polygonal lines in $\mathbb{RP}^n$ and define a notion of "inflection" in terms of n consecutive vertices. It turns out that polygons in $\mathbb{RP}^n$ satisfy an analog of the Barner theorem.

We prove discrete four-vertex and six-vertex theorems for convex plane polygons and a discrete version of Ghys' theorem. The latter concerns a pair of n-tuples of points in $\mathbb{RP}^1$; the Schwarzian derivative is replaced by the cross-ratio of four consecutive points.

Discrete four-vertex and six-vertex theorems

Let P be a plane convex n-gon with $n \geq 4$. Denote the consecutive vertices by $V_1, \ldots, V_n$, where we understand the indices cyclically, that is, $V_{n+i} = V_i$.

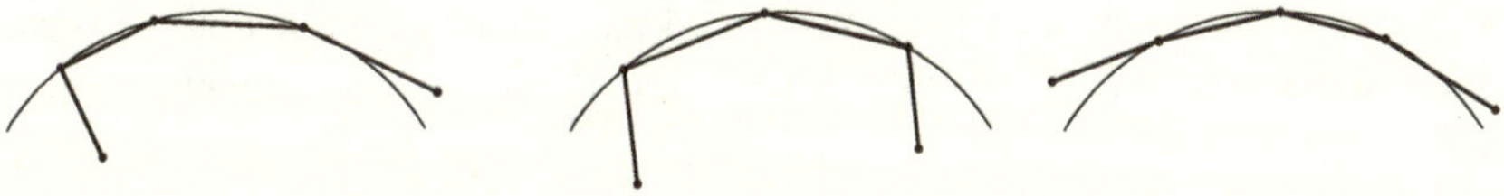

(a) (b)

Fig. 4.8. Triples of vertices: (a) not extremal, (b) extremal

A triple of vertices (V_i, V_{i+1}, V_{i+2}) is called *extremal*[2] if V_{i-1} and V_{i+3} lie on the same side of the circle through V_i, V_{i+1}, V_{i+2} (this does not exclude the case when V_{i-1} or V_{i+3} belongs to the circle); see Figure 4.8.

The following result is an analog of the four-vertex theorem.

Theorem 4.5.1 *Every plane convex polygon P with $n \geq 4$ vertices has at least four extremal triples of vertices.*

Assume now that $n \geq 6$. Similarly to the notion of extremal triples of vertices, we give the following definition. Five consecutive vertices $V_i, \dots, V_{i+4}$ are called *extremal* if V_{i-1} and V_{i+5} lie on the same side of the conic through these five points (this does not exclude the case when V_{i-1} or V_{i+5} belongs to the conic).

A discrete version of the smooth six-vertex theorem is as follows.

Theorem 4.5.2 *Every plane convex polygon P with $n \geq 6$ vertices has at least six extremal quintuples of vertices.*

In spite of being elementary, the next example plays an important role in the proofs.

Example 4.5.3 If P is a quadrilateral or a hexagon, then the respective four- or six-vertex theorem holds tautologically.

As often happens, discretization is not unique. Note here an alternative approach to discretization.

Remark 4.5.4 Consider inscribed circles or conics in consecutive triples or quintuples of sides of a polygon, respectively. Analogs of Theorems 4.5.1 and 4.5.2 hold, providing "dual" theorems. In the case of six vertices, this theorem is equivalent to Theorem 4.5.2 via projective duality. In the case of four vertices,

[2] We have a terminological difficulty here: dealing with polygons, we cannot use the term "vertex" in the same sense as in the smooth case; hence the term "extremal".

both formulations, involving circumscribed and inscribed circles, make sense on the sphere and are equivalent via projective duality therein as well.

The discrete Ghys theorem

A discrete object of study here is a pair of cyclically ordered n-tuples $X = (x_1, \ldots, x_n)$ and $Y = (y_1, \ldots, y_n)$, with $n \geq 4$, of distinct points in $\mathbb{RP}^1$. This is a discretization of a diffeomorphism of $\mathbb{RP}^1$ or, better said, of its graph. Choosing an orientation of $\mathbb{RP}^1$, we assume that the cyclic order of each of the two n-tuples is induced by the orientation.

A triple of consecutive indices $(i, i+1, i+2)$ is called extremal if the difference of cross-ratios

$$[y_j, y_{j+1}, y_{j+2}, y_{j+3}] - [x_j, x_{j+1}, x_{j+2}, x_{j+3}] \tag{4.14}$$

changes sign as j changes from $i-1$ to i (this does not exclude the case when either of the differences vanishes).

Theorem 4.5.5 *For every pair of n-tuples of points X, Y as above there exist at least four extremal triples.*

Example 4.5.6 If $n = 4$ then the theorem holds for the following simple reason. The cyclic permutation of four points induces the next transformation of the cross-ratio:

$$[x_4, x_1, x_2, x_3] = \frac{[x_1, x_2, x_3, x_4]}{[x_1, x_2, x_3, x_4] - 1},$$

which is an involution. Furthermore, if $a > b > 1$ then $a/(a-1) < b/(b-1)$. Therefore, each triple of indices is extremal.

Let us interpret Theorem 4.5.5 in geometrical terms, similarly to Theorems 4.5.1 and 4.5.2. There exists a unique projective transformation that takes points x_i, x_{i+1}, x_{i+2} to y_i, y_{i+1}, y_{i+2}, respectively. The graph γ of this transformation is a hyperbola in $\mathbb{RP}^1 \times \mathbb{RP}^1$ which is an analog of a circle in the four-vertex theorem. The three points $(x_i, y_i), (x_{i+1}, y_{i+1}), (x_{i+2}, y_{i+2})$ lie on γ.

An ordered couple of points (x_j, x_{j+1}) in oriented $\mathbb{RP}^1$ defines the unique segment. An ordered couple of points $((x_j, y_j), (x_{j+1}, y_{j+1}))$ in $\mathbb{RP}^1 \times \mathbb{RP}^1$ also defines the unique segment, the one whose projection on each factor is the defined segment in $\mathbb{RP}^1$.

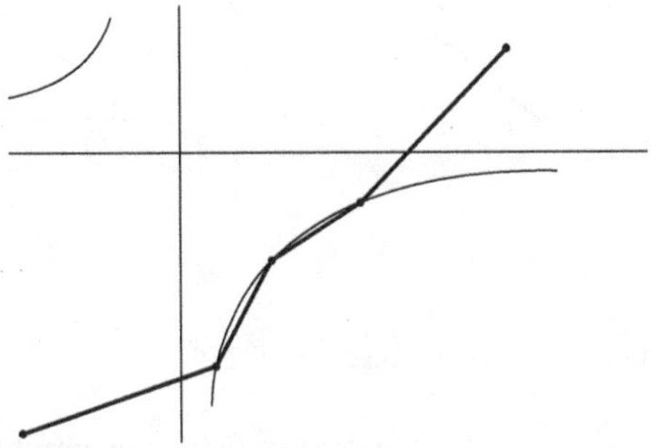

Fig. 4.9. Extremal triple in $\mathbb{RP}^1 \times \mathbb{RP}^1$

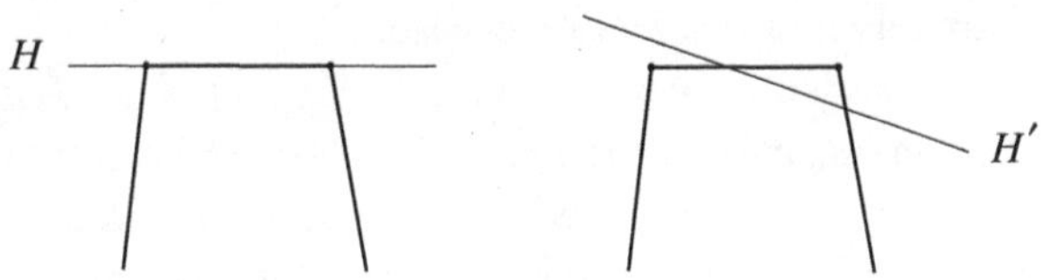

Fig. 4.10. Multiplicity 2

Exercise 4.5.7 The triple $(i, i+1, i+2)$ is extremal if and only if the topological index of intersection of the broken line $(x_{i-1}, y_{i-1}), \ldots, (x_{i+3}, y_{i+3})$ with γ is zero; see Figure 4.9.

Strictly convex polygons in $\mathbb{RP}^n$

Consider a closed polygon P in $\mathbb{RP}^n$, that is, a closed broken line with vertices $V_1, \ldots, V_m$, $m \geq n+1$, in general position. This means that for every set of vertices $V_{i_1}, \ldots, V_{i_k}$, where $k \leq n+1$, the span of $V_{i_1}, \ldots, V_{i_k}$ is $(k-1)$-dimensional.

To deal with polygons in the same manner as with smooth curves, we will need a few technical definitions.

(a) A polygon P is said to be *transverse* to a hyperplane H at a point $X \in P \cap H$ if: (i) X is an interior point of an edge and this edge is transverse to H, or (ii) X is a vertex, the two edges incident to X are transverse to H and are locally separated by H. Clearly, transversality is an open condition.

(b) A polygon P is said to intersect a hyperplane H with *multiplicity* k if, for every hyperplane H', sufficiently close to H and transverse to P, the number of points $P \cap H'$ does not exceed k and, moreover, k is attained for some hyperplane H'; see Figure 4.10.

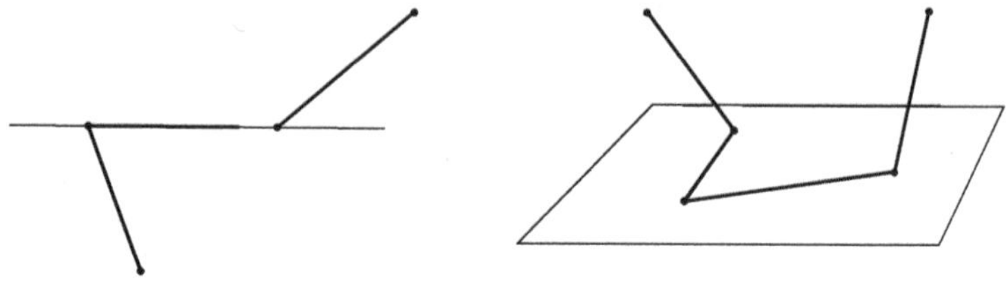

Fig. 4.11. Inflections of polygons in $\mathbb{RP}^2$ and $\mathbb{RP}^3$

(c) Let $\gamma = (A, \ldots, Z)$ be a broken line in $\mathbb{RP}^n$ in general position and let H be a hyperplane not containing A and Z. Denote by k the multiplicity of the intersection of γ with H. We say that A and Z are *separated* by H if k is odd and are *on one side of* H otherwise.

(d) An n-tuple of consecutive vertices $(V_i, \ldots, V_{i+n-1})$ of a polygon P in $\mathbb{RP}^n$ is called an *inflection* if the endpoints V_{i-1} and V_{i+n} of the broken line $(V_{i-1}, \ldots, V_{i+n})$ are separated by the hyperplane through $(V_i, \ldots, V_{i+n-1})$ if n is even, and not separated if n is odd.

The next definition becomes, in the smooth limit, that of strict convexity of smooth curves; see Section 4.3. It allows us to introduce a significant class of polygons which is our main object of study.

A polygon P in $\mathbb{RP}^n$ is called *strictly convex* if through every $n-1$ vertices there passes a hyperplane H such that the multiplicity of its intersection with P equals $n-1$.

Example 4.5.8 Define a simplex $S_n \subset \mathbb{RP}^n$ with vertices $V_1, \ldots, V_{n+1}$ as the projection from the punctured $\mathbb{R}^{n+1}$ to $\mathbb{RP}^n$ of the polygonal line:

$$\widetilde{V}_1 = (1, 0, \ldots, 0),\ \widetilde{V}_2 = (0, 1, 0, \ldots, 0),\ \ldots, \widetilde{V}_{n+1} = (0, \ldots, 0, 1)$$

and

$$\widetilde{V}_{n+2} = (-1)^{n+1}\, \widetilde{V}_1.$$

The last vertex has the same projection as the first one; S_n is contractible for odd n and non-contractible for even n.

We are ready to formulate a discrete version of part (i) of Theorem 4.3.4.

Theorem 4.5.9 *A strictly convex polygon in $\mathbb{RP}^n$ has at least $n+1$ inflections.*

As in the smooth situation, Theorem 4.5.9 implies the three theorems on vertices. The arguments remain intact: the Veronese maps (4.10), (4.11) and the Segre map (4.12) are the same.

Fig. 4.12. Simplexes in $\mathbb{RP}^2$ and $\mathbb{RP}^3$

Proof The proof in the discrete case is significantly simpler than its smooth counterpart. Instead of induction on the dimension of the ambient space, we use induction on the number of vertices m of a polygon.

The base of induction is $m = n + 1$.

Exercise 4.5.10 (a) Prove that the simplex S_n is strictly convex and has $n + 1$ inflections.
(b) Prove that, up to projective transformations, the unique strictly convex $(n + 1)$-gon is the simplex.

Let P be a strictly convex $(m + 1)$-gon with vertices $V_1, \ldots, V_{m+1}$. Delete V_{m+1} and connect V_m with V_1 in such a way that the new edge (V_m, V_1), together with the two deleted ones, (V_m, V_{m+1}) and (V_{m+1}, V_1), form a contractible triangle. Denote the new polygon by P'.

Exercise 4.5.11 Show that P' is strictly convex.

By the induction assumption, the polygon P' has at least $n + 1$ inflections. To complete the proof, it suffices to show that P' cannot have more inflections than P.

A polygon in $\mathbb{RP}^n$ with vertices $V_1, \ldots, V_m$, can be lifted to $\mathbb{R}^{n+1}$ as a polygon with vertices $\widetilde{V}_1, \ldots, \widetilde{V}_m$.

Exercise 4.5.12 Check that an n-tuple $(V_i, \ldots, V_{i+n-1})$ is an inflection if and only if the determinant

$$\Delta_j = \left|\widetilde{V}_j, \ldots, \widetilde{V}_{j+n}\right|$$

changes sign as j changes from $i - 1$ to i. This property is independent of the lifting.

Consider the sequence of determinants $\Delta_1, \Delta_2, \ldots, \Delta_{m+1}$ for the polygon P. Replacing P by P' we remove $n + 1$ consecutive determinants

$$\Delta_{m-n+1}, \Delta_{m-n+2}, \ldots, \Delta_{m+1} \tag{4.15}$$

and add in their place n new determinants

$$\Delta'_{m-n+1}, \Delta'_{m-n+2}, \ldots, \Delta'_m \tag{4.16}$$

where

$$\Delta'_{m-n+i} = \left| \widetilde{V}_{m-n+i} \ldots \widehat{\widetilde{V}}_{m+1} \ldots \widetilde{V}_{m+i+1} \right|$$

with $i = 1, \ldots, n$. The transition from (4.15) to (4.16) is done in two steps. First, we add (4.16) to (4.15) so that the two sequences alternate, that is, we insert Δ'_j between Δ_j and Δ_{j+1}. Second, we delete the "old" determinants (4.15). In the next exercise we prove that the first step preserves the number of sign changes while the second step obviously cannot increase this number.

Exercise 4.5.13 If Δ_{m-n+i} and $\Delta_{m-n+i+1}$ are of the same sign, then Δ'_{m-n+i} is of the same sign too.

These exercises combined yield the proof of Theorem 4.5.9. □

The discrete version of part (ii) is still a conjecture.

Conjecture 4.5.14 *A strictly convex polygon in $\mathbb{RP}^n$ that intersects a hyperplane with multiplicity k has at least k flattenings.*

Comment

This section is based on our work [172]; the reader who is interested in the details of the proofs is invited to consult this paper.

The discretization process itself is worth a discussion. A discrete theorem is a-priori stronger; it becomes, in the limit, a smooth one, thus providing a new proof of the latter. An important feature of the discrete approach is the availability of mathematical induction which can considerably simplify the proofs. Secondly, the very operation of discretization is non-trivial: a single smooth theorem may lead to non-equivalent discrete ones. An example of this phenomenon is provided by recent versions of the four-vertex theorem for convex plane polygons [184, 185, 210, 228].

The discretization approach to projective geometry of curves is not well developed yet. For instance, part (ii) of Theorem 4.3.4 does not have a discrete analog. Another example is a conjectural version of the Möbius theorem: an embedded non-contractible closed polygon in $\mathbb{RP}^2$ has at least three inflections. The number of examples can be easily multiplied.

Interestingly, a certain discrete version of the four-vertex theorem preceded its smooth counterpart by almost hundred years. What we mean is the celebrated Cauchy lemma (1813): *given two convex plane (or spherical) polygons whose respective sides are congruent, the cyclic sequence of the differences of the respective angles of the polygons changes sign at least four times*. This result plays a crucial role in the proof of convex polyhedra rigidity (see [45] for a survey). The Cauchy lemma implies, in the limit, the smooth four-vertex theorem and can be considered as the first result in the area.

4.6 Inflections of Legendrian curves and singularities of wave fronts

In this section we relate global projective geometry of curves to symplectic and contact geometry. The reader is referred to Appendix A.2 for basic notions of symplectic and contact geometry.

We start with the Ghys theorem on zeroes of the Schwarzian derivative and reformulate it in terms of inflections of Legendrian curves in $\mathbb{RP}^3$. We then consider propagation of wave fronts and formulate the Arnold conjecture on Legendrian isotopies. We also discuss here two results on spherical curves: the Segre theorem and the Arnold tennis ball theorem. The section closes with a brief presentation of the curve shortening method proving many theorems discussed in this chapter.

Ghys' theorem and inflections of Legendrian curves

The space $\mathbb{RP}^3$ has a canonical contact structure; see Appendix A.2. The Ghys theorem 4.2.1 has an interesting interpretation in terms of contact geometry.

Consider the linear symplectic space $\mathbb{R}^4$ with the symplectic structure

$$\omega = \mathrm{d}x_1 \wedge \mathrm{d}y_1 - \mathrm{d}x_2 \wedge \mathrm{d}y_2. \tag{4.17}$$

Denote by $\mathbb{RP}^1_1$ and $\mathbb{RP}^1_2$ the projectivizations of the symplectic subspaces $\mathbb{R}^2_1$ with coordinates (x_1, y_1) and $\mathbb{R}^2_2$ with coordinates (x_2, y_2), respectively.

Theorem 4.6.1 *Let $\gamma \subset \mathbb{RP}^3$ be a closed Legendrian curve such that its projections on $\mathbb{RP}^1_1$ from $\mathbb{RP}^1_2$ and on $\mathbb{RP}^1_2$ from $\mathbb{RP}^1_1$ are diffeomorphisms. Then γ has at least four distinct inflection points.*

Proof The curve γ defines a diffeomorphism from $\mathbb{RP}^1_1$ to $\mathbb{RP}^1_2$ as the composition of the two projections; see Figure 4.13.

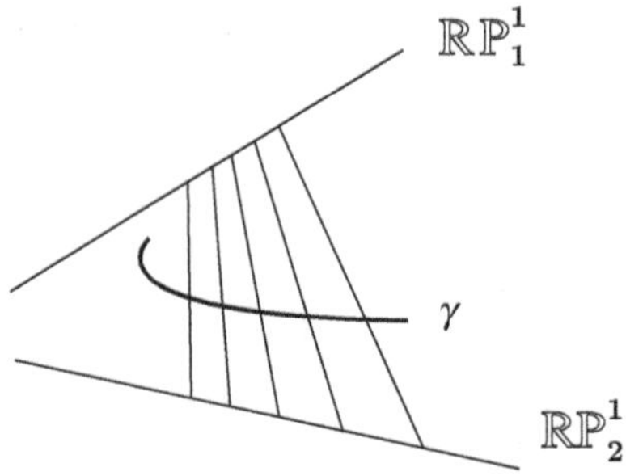

Fig. 4.13. Legendrian graph in $\mathbb{RP}^3$

The converse holds as well. Given a diffeomorphism $f : \mathbb{RP}^1_1 \to \mathbb{RP}^1_2$, there exists a unique area-preserving homogeneous (of degree 1) diffeomorphism $F : \mathbb{R}^2_1 \setminus \{0\} \to \mathbb{R}^2_2 \setminus \{0\}$ that projects to f.

Exercise 4.6.2 Show that the diffeomorphism F is given in polar coordinates by

$$F(\alpha, r) = \left(f(\alpha),\ r \left(\frac{\mathrm{d}f}{\mathrm{d}\alpha} \right)^{-1/2} \right). \tag{4.18}$$

The graph gr$F$ is a Lagrangian submanifold in $\mathbb{R}^2_1 \oplus \mathbb{R}^2_2$ with the symplectic form (4.17) and the projectivization of gr$F$ is a Legendrian curve. If f is projective, then F is linear and the projectivization γ of gr$F$ is a Legendrian line in $\mathbb{RP}^3$.

To complete the proof, it suffices to show that the points in which f is abnormally well approximated by a projective transformation correspond to inflections of γ. Indeed, at such points, γ has a three-point contact with a Legendrian line.

The result now follows from the Ghys theorem. □

Seemingly, we proved more than Theorem 4.6.1 asserts: we found not only inflections of γ but three-point tangencies of γ with its tangent line. In fact, for Legendrian curves, the two coincide.

Lemma 4.6.3 *The inflection points of a Legendrian curve in $\mathbb{RP}^3$ are the points of second-order contact with a line.*

Proof Consider a parameterization $\Gamma(t)$ of a lift of γ to $\mathbb{R}^4$. We need to prove that if the vectors $\Gamma, \dot{\Gamma}, \ddot{\Gamma}, \dddot{\Gamma}$ are linearly dependent then so are $\Gamma, \dot{\Gamma}, \ddot{\Gamma}$.

Since Γ is Legendrian, one has $\omega(\Gamma, \dot{\Gamma}) = 0$. Differentiate to obtain: $\omega(\Gamma, \ddot{\Gamma}) = 0$. This means that the projection $\ddot{\gamma}$ belongs to the contact plane, that is, the contact plane osculates a Legendrian curve. Differentiate once again: $\omega(\dot{\Gamma}, \ddot{\Gamma}) + \omega(\Gamma, \dddot{\Gamma}) = 0$. The inflection condition reads: $(\omega \wedge \omega)(\Gamma, \dot{\Gamma}, \ddot{\Gamma}, \dddot{\Gamma}) = 0$. In view of the two previous formulæ

$$0 = (\omega \wedge \omega)(\Gamma, \dot{\Gamma}, \ddot{\Gamma}, \dddot{\Gamma}) = \omega(\Gamma, \dddot{\Gamma})\, \omega(\dot{\Gamma}, \ddot{\Gamma}) = -\omega(\dot{\Gamma}, \ddot{\Gamma})^2.$$

Thus all three vectors $\Gamma, \dot{\Gamma}, \ddot{\Gamma}$ are pairwise orthogonal with respect to ω. Since ω is non-degenerate, they are linearly dependent. □

Remark 4.6.4 A generic smooth curve in $\mathbb{RP}^3$ never has a three-point contact with a straight line. We have seen that for generic Legendrian curves this happens in isolated points. Moreover, a three-point contact of a Legendrian curve with a Legendrian line is automatically a four-point contact.

We believe that Theorem 4.6.1 can be significantly improved.

Conjecture 4.6.5 *If a Legendrian curve $\gamma \subset \mathbb{RP}^3$ is isotopic to a Legendrian line in the class of embedded Legendrian curves then γ has at least four distinct inflection points.*

Singularities of wave fronts and Legendrian isotopies

The four-vertex theorem can be interpreted in terms of singularities of wave fronts.

Let γ be a plane oval. Consider a wave, propagating with the unit speed from γ in the inward direction, that is, the one-parameter family of curves, equidistant from γ. At the beginning, the equidistant curves remain smooth, but eventually they develop singularities, generically semi-cubic cusps; see Figure 4.14.

Exercise 4.6.6 The locus of singular points of all the equidistant curves is the caustic of γ.

The topology of equidistant curves changes when the equidistant curve passes through a singularity of the caustic: at such a point, the number of cusps changes by 2. The four-vertex theorem implies that some intermediate equidistant curves have four cusps.

The equidistant curves are cooriented by the direction of evolution. An important property of wave propagation is that at no time is an equidistant curve

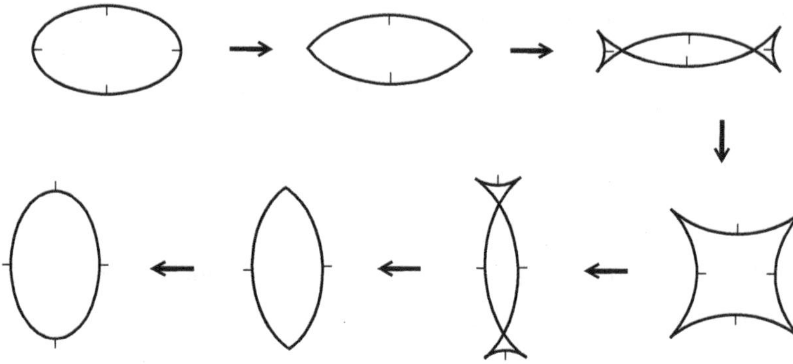

Fig. 4.14. Singularities of equidistants

tangent to itself with coinciding coorientation. This follows from the Huygens principle of classical mechanics (see [10]).

More conceptually, let us lift all equidistant curves to the space of cooriented contact elements of the plane. We obtain a one-parameter family of smooth Legendrian curves whose fronts are the equidistant curves. The absence of self tangencies means that this family is a Legendrian isotopy.

The above discussion leads to the following conjecture.

Conjecture 4.6.7 *Let γ_0 and γ_1 be plane ovals, cooriented inward and outward, respectively, and let Γ_0 and Γ_1 be their Legendrian lifts to the space of cooriented contact elements. Then for every Legendrian isotopy Γ_t, there exists a value of t for which the front γ_t has at least four cusps.*

Segre's theorem and the tennis ball theorem

Consider a smooth simple closed curve γ on S^2. The *convex hull* of γ is a convex body, equal to the intersection of all closed half-spaces containing γ. Let O be a point inside the convex hull of γ. If $O \in \gamma$, assume that O is not a vertex.

Theorem 4.6.8 *There are at least four distinct points of γ at which the osculating plane passes through O.*

The known proofs of this theorem are technically complicated and we do not reproduce them here. The next special case is particularly attractive.

Theorem 4.6.9 *If γ bisects the area of the sphere then it has at least four distinct points of contact of order 2 (a three-point contact) with a great circle.*

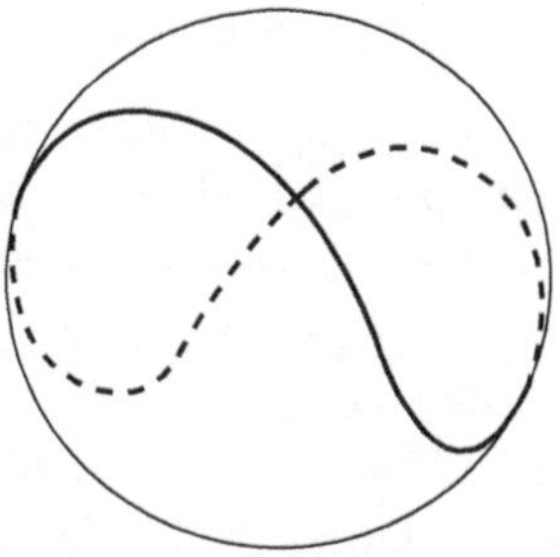

Fig. 4.15. Tennis ball

This theorem is called the "tennis ball theorem"; see Figure 4.15. It resembles the Möbius theorem; see Section 4.1. The points of order 2 contact with a great circle are the points where the geodesic curvature of the curve vanishes, that is, the inflection points of the spherical curve.

The tennis ball theorem is a particular case of Theorem 4.6.8. Indeed, since γ bisects the area of S^2, it intersects every great circle. This implies that the center of S^2 lies in the convex hull of γ.

Proofs by curve shortening

An interesting approach to the tennis ball theorem, along with Theorems 4.1.1, 4.1.10 and the Möbius theorem, consists in studying the curve shortening flow. This flow is defined as

$$\frac{\partial \gamma_s}{\partial s} = \kappa \nu \tag{4.19}$$

where $\gamma_s(t)$ is a one-parameter family of curves parameterized by the arc-length t, the function $\kappa = \kappa_s(t)$ is the curvature of the curve and $\nu = \nu_s(t)$ is the unit normal vector. In the affine case, t is the affine arc-length, κ is the affine curvature and $\kappa\nu = \partial^2\gamma_s/\partial t^2$ is the affine normal vector. The equation (4.19) can be regarded as the heat equation.

The main idea of the proofs is as follows. One proves that the flow is defined for all times, that is, γ_s remains smooth and simple. For the tennis ball theorem, the area-bisecting property persists as well. By an appropriate version of the maximum principle, the number of the points of γ under consideration (vertices, sextactic points or inflections) does not increase with time. Next, one analyses the limit shape of the curve γ_s for large s: the circle (after rescaling) in the Euclidean case, an ellipse (also after rescaling) in the affine case, a straight line in the Möbius case, a great circle in the tennis ball case. For sufficiently large s, one concludes by an appropriate version of the Sturm theorem; see Appendix A.1.

Comment

The symplectic and contact viewpoint on Sturm theory and global geometry of curves was developed and popularized by V. Arnold; see [11]–[14] and references therein. He discovered intimate relations between the four-vertex theorem and symplectic and contact topology, in particular, Legendrian knot theory. Arnold's work stimulated recent interest in the subject.

Theorem 4.6.1 and Conjecture 4.6.5 can be found in [171]. This conjecture is similar to Conjecture 4.6.7, due to Arnold, a positive solution of which was recently announced by Yu. Chekanov and P. Pushkar'. Classic Theorem 4.6.8 is due to Segre [188], see also [229] for a simpler proof, Theorem 4.6.9 was proved in [11] in a different way. Curve shortening problems have been thoroughly studied in the recent years; see [42]. Our exposition follows [6].

5

Projective invariants of submanifolds

This chapter concerns projective geometry and projective topology of submanifolds of dimension greater than 1 in projective space. We start with a panorama of classical results concerning surfaces in $\mathbb{RP}^3$. This is a thoroughly studied subject, and we discuss only selected topics connected with the main themes of this book. Section 5.2 concerns relative, affine and projective differential geometry of non-degenerate hypersurfaces. In particular, we construct a projective differential invariant of such a hypersurface, a $(1, 2)$-tensor field on it. Section 5.3 is devoted to various geometrical and topological properties of a class of transverse fields of directions along non-degenerate hypersurfaces in affine and projective space, the exact transverse line fields. In Section 5.4 we use these results to give a new proof of a classical theorem: the complete integrability of the geodesic flow on the ellipsoid and of the billiard inside the ellipsoid. Section 5.5 concerns Hilbert's fourth problem: to describe Finsler metrics in a convex domain in projective space whose geodesics are straight lines. We describe integral-geometric and analytic solutions to this celebrated problem in dimension 2 and discuss the multi-dimensional case as well. The last section is devoted to Carathéodory's conjecture on two umbilic points on an ovaloid and recent conjectures of Arnold on global geometry and topology of non-degenerate closed hypersurfaces in projective space.

Once again, our account is far from being comprehensive. The choice of material reflects our interests and tastes; we also tried to combine older classic results with newer and lesser known ones. Many things are not even mentioned; a notable example is the theory of Cartan connections, in particular, projective connections, and its applications in projective differential geometry.

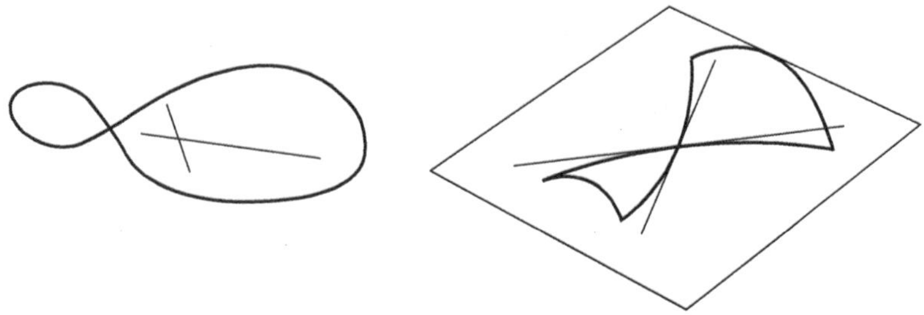

Fig. 5.1. Non-degenerate "hyperbolic" surface: asymptotic directions

5.1 Surfaces in $\mathbb{RP}^3$: differential invariants and local geometry

Geometry of surfaces in the projective space is a classical and thoroughly studied subject; there are about a dozen books devoted to it. In this section we give a concise exposition including this material into our general framework: differential operators on tensor densities, action of diffeomorphism groups and the notion of projective structure on curves. When possible, we emphasize similarities with the case of curves in the projective plane.

We will consider *non-degenerate* surfaces. This means that, in an affine coordinate system, the second quadratic form of the surface is non-degenerate at every point; this condition is independent of the choice of affine coordinates. Moreover, following a well-established tradition, we will assume that the second quadratic form has signature (1, 1), that is, the surface is saddle-shaped. The standard torus $x_1 x_2 = x_0 x_3$ is an example of such a surface; in affine coordinates this corresponds to a hyperbolic paraboloid.

The other, "elliptic" case will not be covered here. In the classic period of projective differential geometry all objects were analytic and could be continued to the complex domain, so that the sign assumptions did not affect the formulæ. In the smooth case, the difference is substantial and we are not aware of a natural geometric theory of locally convex projective surfaces.

Wilczynski parameterizations

Let $M^2 \subset \mathbb{RP}^3$ be a non-degenerate saddle-shaped surfaces. At every point $x \in M$, one has two distinguished tangent lines, called the *asymptotic directions*; these lines are tangent to the intersection curves of M with the tangent plane $T_x M$. This defines two transverse fields of directions on M and therefore two one-dimensional foliations. One can choose local coordinates (u, v) such that the leaves of the foliations are given by ($u =$ const, $v =$ const). These coordinates are called asymptotic; they are defined up to reparameterization $(u, v) \mapsto (U(u), V(v))$. Thus the local action of the group $\mathrm{Diff}(\mathbb{R}^2)$ is replaced by that of a much smaller group $\mathrm{Diff}(\mathbb{R}) \times \mathrm{Diff}(\mathbb{R})$.

We consider a parameterized surface $x(u, v) \subset \mathbb{RP}^3$. Let $X(u, v) \subset \mathbb{R}^4$ be its arbitrary lift.

Exercise 5.1.1 Show that the four vectors X, X_u, X_v, X_{uv} are linearly independent for every (u, v).

By analogy with the canonical lift of projective curves, one can uniquely fix the lift of the parameterized surface $x(u, v)$ into $\mathbb{R}^4$ by the condition

$$|X\, X_u\, X_v\, X_{uv}| = 1. \tag{5.1}$$

Let us call this lift *canonical.*

Theorem 5.1.2 *(i) The coordinates of the canonical lift satisfy the system of linear differential equations*

$$\begin{aligned} X_{uu} + b\, X_v + f\, X &= 0, \\ X_{vv} + a\, X_u + g\, X &= 0, \end{aligned} \tag{5.2}$$

where a, b, f, g are functions in (u, v) satisfying the integrability conditions

$$\begin{aligned} f_{vv} - 2b_v\, g - b\, g_v - g_{uu} + f_u\, a + 2a_u\, f &= 0, \\ b\, a_v + 2b_v\, a + a_{uu} + 2g_u &= 0, \\ a\, b_u + 2a_u\, b + b_{vv} + 2f_v &= 0. \end{aligned} \tag{5.3}$$

(ii) Conversely, system (5.2) whose coefficients satisfy relations (5.3) corresponds to a non-degenerate parameterized surface $M \subset \mathbb{RP}^3$.

Proof Let $X(u, v)$ be the canonical lift of a surface $x(u, v) \subset \mathbb{RP}^3$. Since the vectors X, X_u, X_v, X_{uv} are linearly independent, X_{uu} and X_{vv} are their linear combinations. Since (u, v) are asymptotic coordinates, these linear combinations do not contain X_{uv}. Indeed, the preimage of the tangent plane $T_x M$ in $\mathbb{R}^4$ is the linear span of X, X_u, X_v. The second derivative measures the deviation of the surface from its tangent plane, but the tangent line to an asymptotic curve is tangent to the surface with order 2.

Therefore the linear combinations are of the form

$$\begin{aligned} X_{uu} + b\, X_v + c\, X_u + f\, X &= 0 \\ X_{vv} + a\, X_u + d\, X_v + g\, X &= 0 \end{aligned} \tag{5.4}$$

Differentiating (5.1) with respect to u, one obtains

$$|X\, X_{uu}\, X_v\, X_{uv}| + |X\, X_u\, X_v\, X_{uuv}| = 0.$$

Substitute the above equation for X_{uu} into the first determinant and the derivative $(X_{uu})_v$ into the second determinant. Each determinant turns out to be equal to c. Hence $c = 0$ and likewise for d. This proves system (5.2).

Exercise 5.1.3 Assuming that $X(u, v)$ satisfies (5.2), check that the identity $(X_{uu})_{vv} = (X_{vv})_{uu}$ is equivalent to system (5.3).

Let us prove part (ii). In a standard way, one replaces the system of second-order equations (5.2) by a (matrix) system of first-order equations. Namely, one introduces a new variable: $\mathcal{X} = (X, X_u, X_v, X_{uv})$, and then system (5.2) can be rewritten as

$$\left(\frac{\partial}{\partial u} + A\right)\mathcal{X} = 0, \qquad \left(\frac{\partial}{\partial v} + B\right)\mathcal{X} = 0$$

where the linear operators A and B are easily computed in terms of a, b, f, g.

Exercise 5.1.4 Check that system (5.3) is equivalent to the equation $[\partial/\partial u + A, \partial/\partial v + B] = 0$.

To summarize, systems (5.2) and (5.3) together define a flat connection in the trivial bundle over the coordinate domain (u, v) with fiber $\mathcal{X}$. Horizontal sections of this connection are the desired solutions. □

System (5.2) is called the canonical (or the Wilczynski) system of differential equations associated with a surface in $\mathbb{RP}^3$. It plays the same role as the Sturm–Liouville equation in the case of projective structures, or the operator (2.7) in the case of non-degenerate curves in $\mathbb{RP}^n$.

Affine and Fubini parameterizations

The general system (5.4) describes an arbitrary lift of M to $\mathbb{R}^4$.

Exercise 5.1.5 Show that the compatibility condition $(X_{uu})_{vv} = (X_{vv})_{uu}$ implies that $c_v = d_u$.

Hence (locally) one has $c = h_u$ and $d = h_v$ where $h(u, v)$ is a function.

We will consider two more lifts of M. The first one is the *affine lift*. This lift is characterized by the condition that $X(u, v)$ lies in an affine hyperplane.

Lemma 5.1.6 *The affine lift of a surface $M \subset \mathbb{RP}^3$ satisfies the system of equations*

$$\begin{aligned} X_{uu} + b\, X_v + h_u\, X_u &= 0, \\ X_{vv} + a\, X_u + h_v\, X_v &= 0. \end{aligned} \tag{5.5}$$

Proof For the affine lift, one of the components of $X(u, v)$ is a constant, and this component satisfies system (5.4). Hence $f = g = 0$. □

The coefficients a, b and h satisfy a system of two non-linear equations of compatibility; we do not dwell on this.

For historical reasons, we mention the Fubini lift characterized by the equation $e^h = ab$ in system (5.4).

Tensor densities and solutions of the fundamental system

Our approach is similar to that of Section 2.4. To start with, we will explain the geometric, tensor meaning of the solutions of the canonical system (5.2).

The group $\mathrm{Diff}(\mathbb{R}) \times \mathrm{Diff}(\mathbb{R})$ locally acts on $M \subset \mathbb{RP}^3$, preserving the asymptotic foliations. We will determine the transformation law for the canonical system (5.2) with respect to this action.

Lemma 5.1.7 *The solutions of system (5.2) are tensor densities of the form*

$$X = X(u, v)\, (\mathrm{d}u)^{-\frac{1}{2}} (\mathrm{d}v)^{-\frac{1}{2}}. \tag{5.6}$$

Proof Let $(u, v) \mapsto (\varphi(u), \psi(v))$ be a local diffeomorphism of M. This diffeomorphism acts on the parameterized surface $x(u, v)$ in the usual way $x(u, v) \mapsto x(\varphi^{-1}(u), \psi^{-1}(v))$. We wish to compute how this affects the canonical lift.

Let us find explicitly the canonical lift of M to $\mathbb{R}^4$. In affine coordinates on $\mathbb{RP}^3$, one has

$$x(u, v) = (x_1(u, v) : x_2(u, v) : x_3(u, v) : 1)\,.$$

Then the canonical lift to $\mathbb{R}^4$ is of the form

$$X(u, v) = (r\, x_1, r\, x_2, r\, x_3, r)$$

where $r(u, v)$ is a non-vanishing function. Condition (5.1) implies

$$r(u, v) = \begin{vmatrix} x_{1u} & x_{2u} & x_{3u} \\ x_{1v} & x_{2v} & x_{3v} \\ x_{1uv} & x_{2uv} & x_{3uv} \end{vmatrix}^{-\frac{1}{4}}$$

It remains to apply the Chain Rule to obtain the transformation law

$$\left(\varphi^{-1} \times \psi^{-1}\right)^* : r(u, v) \mapsto \left(\varphi'(u)\right)^{-1/2} \left(\psi'(v)\right)^{-1/2} r(\varphi(u), \psi(v)).$$

Hence the result. □

Generalizing the last formula, let us introduce the space $\mathcal{F}_{\lambda,\mu}(M)$ of *polarized tensor densities* on the surface $M \subset \mathbb{RP}^3$. In fact, we need only a transverse pair of foliations[1] on M. The tangent bundle of M is split: $TM = L_1 \oplus L_2$ where L_1 and L_2 are the tangent bundles of the foliations. The space $\mathcal{F}_{\lambda,\mu}(M)$ is the space of sections of the line bundle $(L_1^*)^{\otimes\lambda} \otimes (L_2^*)^{\otimes\mu}$.

In local coordinates, a (λ, μ)-density is given by

$$\rho = r(u, v)\,(\mathrm{d}u)^\lambda (\mathrm{d}v)^\mu,$$

and the local action of the group $\mathrm{Diff}(\mathbb{R}) \times \mathrm{Diff}(\mathbb{R})$ is written as

$$\mathrm{T}^{\lambda,\mu}_{(\varphi^{-1}\times\psi^{-1})} : r(u, v) \mapsto \left(\varphi'(u)\right)^\lambda \left(\psi'(v)\right)^\mu r(\varphi(u), \psi(v)). \tag{5.7}$$

Differential invariants

We will now investigate how $\mathrm{Diff}(\mathbb{R}) \times \mathrm{Diff}(\mathbb{R})$ acts on the coefficients of the canonical system. In other words, we wish to see which equation is satisfied by the transformed solutions.

Proposition 5.1.8 *(i) The coefficients a and b of the fundamental system (5.2) have the meaning of tensor densities of degrees* $(-1, 2)$ *and* $(2, -1)$ *respectively, that is,*

$$\alpha = a(u, v)\,(\mathrm{d}u)^{-1}(\mathrm{d}v)^2, \qquad \beta = b(u, v)\,(\mathrm{d}u)^2(\mathrm{d}v)^{-1} \tag{5.8}$$

are well-defined elements of $\mathcal{F}_{-1,2}(M)$ *and* $\mathcal{F}_{2,-1}(M)$.
(ii) The Sturm–Liouville operators

$$\frac{\mathrm{d}^2}{\mathrm{d}u^2} + \left(f + \frac{1}{2} b_v\right), \qquad \frac{\mathrm{d}^2}{\mathrm{d}v^2} + \left(g + \frac{1}{2} a_u\right), \tag{5.9}$$

[1] Called a 2-web; web geometry is a well-developed and rich subfield of differential geometry.

the first depending on v and the second on u as a parameter, are well-defined.

Proof Consider the transformed solution $\bar{X} = \mathrm{T}^{-1/2,-1/2}_{(\varphi^{-1}\times\psi^{-1})} X$.

Exercise 5.1.9 Check that $\bar{X}$ satisfies system (5.2) with the new coefficients

$$\bar{a} = \frac{(\psi')^2}{\varphi'} a(\varphi, \psi), \qquad \bar{b} = \frac{(\varphi')^2}{\psi'} b(\varphi, \psi)$$

$$\bar{f} = (\varphi')^2 f(\varphi, \psi) + \frac{1}{2} S(\varphi) + \frac{\psi'' (\varphi')^2}{2(\psi')^2} b(\varphi, \psi)$$

$$\bar{g} = (\psi')^2 g(\varphi, \psi) + \frac{1}{2} S(\psi) + \frac{\varphi'' (\psi')^2}{2(\varphi')^2} a(\varphi, \psi)$$

The transformation law for the coefficients a and b proves (5.8). On the other hand, it follows from the transformation law for the coefficients f and g that the quantities $f + b_v/2$ and $g + a_u/2$ transform as the potentials of the Sturm–Liouville operator, see (1.12). □

Proposition 5.1.8, part (i), provides two differential invariants (5.8) of a surface $M \subset \mathbb{RP}^3$. Traditionally, one considers their sum $\alpha + \beta$, called the "linear projective element", their product which is a quadratic form and a cubic form, namely

$$\alpha\beta = ab\,\mathrm{d}u\,\mathrm{d}v, \qquad \alpha\beta^2 + \alpha^2\beta = ab^2\,\mathrm{d}u^3 + a^2b\,\mathrm{d}v^3. \tag{5.10}$$

Part (ii) of Proposition 5.1.8 means that each asymptotic curve has a canonical projective structure.

Although the quadratic form $\alpha\beta$ is a weaker invariant than α and β separately, it allows us to develop local differential geometry on M. One can associate with this form a Levi–Civita connection and an area form; the geodesics of this connection are often called the projective geodesics.

Remark 5.1.10 In the classic period of projective differential geometry, the main problem was to describe a complete set of projective invariants (of curves, surfaces, etc.). These invariants should be tensor fields. In our situation, one can construct two invariants $h \in \mathcal{F}_{2,0}(M)$ and $k \in \mathcal{F}_{0,2}(M)$ given by

$$h = f + \frac{1}{2} b_v - \frac{1}{16}\frac{b_{uu}}{b} + \frac{5}{64}\frac{b_u^2}{b^2}, \qquad k = g + \frac{1}{2} a_u - \frac{1}{16}\frac{a_{vv}}{a} + \frac{5}{64}\frac{a_u^2}{a^2}$$

which, along with a and b, form the complete system of invariants of M; cf. Remark 2.4.5.

Commuting differential operators

Another way to understand the canonical system of equations (5.2) and the integrability conditions (5.3) is to construct two families of commuting linear differential operators.

System (5.2) can be written in terms of two differential operators:

$$A : \mathcal{F}_{-1/2,-1/2}(M) \to \mathcal{F}_{-1/2,3/2}(M), \qquad B : \mathcal{F}_{-1/2,-1/2}(M) \to \mathcal{F}_{3/2,-1/2}(M)$$

where

$$A = \frac{\partial^2}{\partial v^2} + a\,\frac{\partial}{\partial u} + g, \qquad B = \frac{\partial^2}{\partial u^2} + b\,\frac{\partial}{\partial v} + f,$$

namely, $A(X) = 0$ and $B(X) = 0$. We have seen that the "modified" coefficients

$$\widehat{f} = f + \frac{1}{2}\,b_v, \qquad \widehat{g} = g + \frac{1}{2}\,a_u$$

play a geometric role as the potentials of the induced projective structures on the asymptotic curves; see Proposition 5.1.8.

Let us introduce two families of differential operators

$$A_\lambda : \mathcal{F}_{\lambda,-1/2}(M) \to \mathcal{F}_{\lambda,3/2}(M), \qquad B_\mu : \mathcal{F}_{-1/2,\mu}(M) \to \mathcal{F}_{3/2,\mu}(M)$$

generalizing the operators A and B. They are given by the formulæ

$$A_\lambda = \frac{\partial^2}{\partial v^2} + a\,\frac{\partial}{\partial u} + \widehat{g} + \lambda\,a_u, \qquad B_\mu = \frac{\partial^2}{\partial u^2} + b\,\frac{\partial}{\partial v} + \widehat{f} + \mu\,b_v. \quad (5.11)$$

Obviously, $A_{-1/2} = A$ and $B_{-1/2} = B$.

Exercise 5.1.11 Check that the coefficients of the operators A_λ and B_μ transform under the action of $\mathrm{Diff}(\mathbb{R}) \times \mathrm{Diff}(\mathbb{R})$ in exactly the same way as those of A and B.

The following result is verified by a direct computation.

Proposition 5.1.12 *Conditions (5.3) are equivalent to the equation*

$$A_{3/2} \circ B_{-1/2} - B_{3/2} \circ A_{-1/2} = 0. \quad (5.12)$$

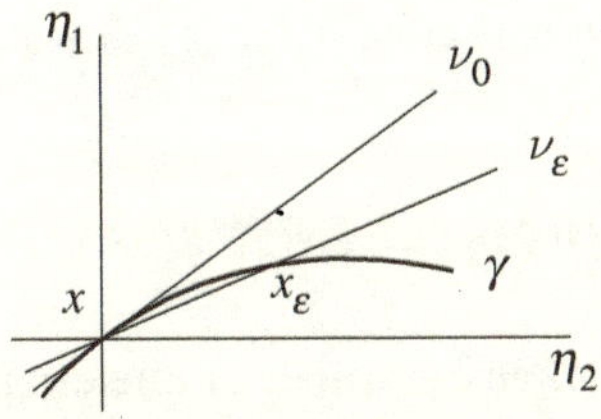

Fig. 5.2. Bompiani construction

Geometric construction for α and β (Bompiani)

Differential invariants α and β have geometric interpretation in terms of cross-ratio.

Consider a plane curve γ and let η_1, η_2 be two smooth fields of directions along γ. We will construct a differential 1-form λ on γ. Let x be a point of γ and ξ a tangent vector to γ at x. Extend ξ to a tangent vector field in a vicinity of x and denote by x_ε the time-ε image of x in the respective flow. Let ν_0 be the tangent line to γ at x and ν_ε be the line (x, x_ε); see Figure 5.2.

Exercise 5.1.13 (a) Check that the cross-ratio satisfies

$$[\nu_0, \eta_1(x), \eta_2(x), \nu_\varepsilon] = 1 + \frac{\varepsilon}{2}\lambda(\xi) + O(\varepsilon^2)$$

where $\lambda(\xi)$ is a linear function of ξ.
(b) Let $\gamma(t)$ be a parameterization of the curve γ such that $\gamma(0) = x$, $\gamma'(0) = \xi$, and let ξ_1, ξ_2 be vectors along the lines $\eta_1(x)$ and $\eta_2(x)$, respectively. Check that

$$\lambda(\xi) = \frac{\gamma''(0) \times \xi_1}{\gamma'(0) \times \xi_1} - \frac{\gamma''(0) \times \xi_2}{\gamma'(0) \times \xi_2} \tag{5.13}$$

where $\times$ denotes the cross-product of vectors in the plane.

We will use the 1-form λ to give a geometric interpretation of the differential invariants α and β.

Pick a point $x \in M$ and consider the intersection of the tangent plane $T_x M$ with M. Generically, this intersection is a curve that has a normal double point at x. Let γ be the branch of the intersection curve, tangent to the v-asymptotic direction. Choose η_1 to be the u-asymptotic direction and η_2 arbitrarily. Fix a vector ξ tangent to γ and vectors ξ_1, ξ_2 along η_1, η_2.

The differential invariant $\alpha \in \mathcal{F}_{-1,2}(M)$ is a function with one argument a tangent vector to the u-asymptotic direction and two arguments tangent vectors to the v-asymptotic direction, that is, $\alpha(\xi_1, \xi_2^v, \xi)$ where ξ_2^v is the v-component of ξ_2.

Exercise 5.1.14 Check that $\lambda(\xi) = \frac{2}{3}\,\alpha(\xi_1, \xi_2^v, \xi)$.

Hint. It is convenient to work in the affine parameterization (5.5). Give the curve a parameterization and use formula (5.13). To find the u-component $\gamma''(0)^u$, use the fact that $\gamma'''(0)$ lies in the tangent plane.

A similar construction applies to the invariant β and, as a consequence, we obtain geometric interpretations of the quadratic and cubic forms of the surface.

Geometric construction for the quadratic form

We will describe a different geometric construction that allows us to compute the quadratic form $\alpha\beta$ in terms of cross-ratio.

Pick a point $x \in M$ and vectors ξ_1, ξ_2, tangent to the u and v-asymptotic directions, respectively. Our goal is to determine $(\alpha\beta)(\xi_1, \xi_2)$. Choose coordinates (u, v) in such a way that x is the origin and ξ_1, ξ_2 are the unit coordinate tangent vectors. Consider the points $(\varepsilon, 0)$ and $(0, \varepsilon)$. Let η_1 and η_2 be the u- and v-asymptotic tangent lines at the origin and η_1^ε and η_2^ε the u- and v-asymptotic tangent lines at the point $(\varepsilon, 0)$ and $(0, \varepsilon)$ respectively; see Figure 5.3. In spite of the fact that these four lines are neither coplanar nor concurrent, we will define their cross-ratio.

Consider an affine lift of M and choose a parallel projection of the ambient affine space onto the tangent plane at x. We obtain four lines in T_xM. Parallel translate η_1^ε and η_2^ε to the origin.

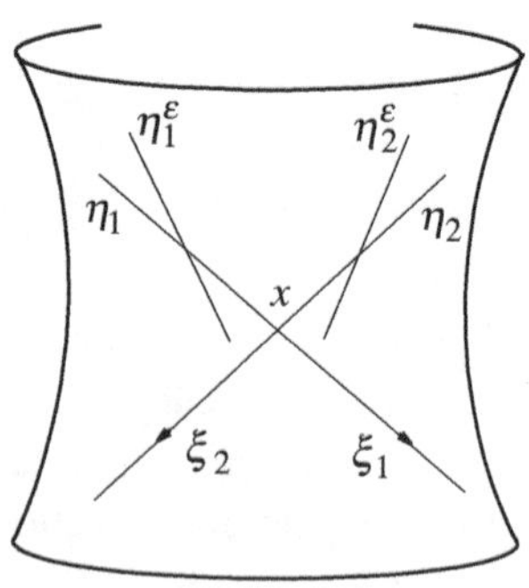

Fig. 5.3. Construction of the quadratic form

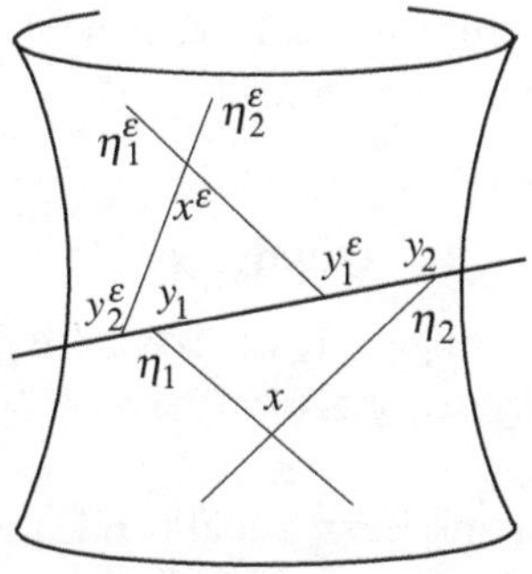

Fig. 5.4. Construction of the linear projective element

Exercise 5.1.15 Check that the cross-ratio of the four concurrent lines is

$$[\eta_1, \eta_2, \eta_1^\varepsilon, \eta_2^\varepsilon] = -\varepsilon^2 ab + O(\varepsilon^3).$$

The construction is independent of the choice of an affine lift and a parallel projection since the result involves only α and β, which are projective invariants.

Geometric construction for the linear projective element

Fix a point $x \in M$ and let $\xi = (\xi^u, \xi^v)$ be a tangent vector at x. As before, consider the point $x_\varepsilon \in M$ obtained from x moving distance ε in the direction of ξ. Let η_1 and η_2 be the u- and v-asymptotic tangent lines at x and η_1^ε and η_2^ε the respective asymptotic tangent lines at x_ε. Denote by ℓ the intersection line of the (projective) tangent planes $T_x M$ and $T_{x_\varepsilon} M$ and by $y_1, y_2, y_1^\varepsilon, y_2^\varepsilon$ the intersection points of ℓ with the respective four asymptotic lines; see Figure 5.4.

Exercise 5.1.16 Check that the cross-ratio of the four points satisfies

$$[y_1, y_1^\varepsilon, y_2^\varepsilon, y_2] = \varepsilon^2 \left(\frac{1}{2}(\alpha + \beta)(\xi)\right)^2 + O(\varepsilon^3)$$

where $\alpha(\xi) = a(u, v)\,(\xi^u)^2/\xi^v$ and similarly for β.

Hint. It is convenient to make computations in the canonical lift.

Note that the construction makes sense for a non-degenerate surface since the tangent planes at x and x_ε do not coincide. Note also that the cross-ratio is identically zero for a quadratic surface. Indeed, in this case, the four lines lie on M and the lines η_1 and η_1^ε intersect η_2 and η_2^ε. Therefore $y_2 = y_1^\varepsilon$ and $y_1 = y_2^\varepsilon$.

Quadrics

A non-degenerate quadric of signature (2, 2) is a prime example of a projective surface in $\mathbb{RP}^3$. We consider quadrics from the point of view of the fundamental system (5.2).

Recall two classic theorems. First, a doubly ruled surface in $\mathbb{R}^3$ is a quadric. Second, given three pairwise skew lines in $\mathbb{RP}^3$, the one-parameter family of lines that intersect all three lie on a non-degenerate quadric; see, for example, [22].

All quadrics of signature (2, 2) are projectively equivalent to the standard torus. It can be parameterized as follows

$$x(u, v) = (1 : u : v : uv). \tag{5.14}$$

Lemma 5.1.17 *A surface described by system (5.2) is a quadric if and only if $a(u, v) = b(u, v) = 0$ everywhere.*

Proof Let M be a quadric, then the asymptotic tangent lines lie entirely on M. It follows from the Bompiani construction of the differential invariants α and β that these invariants identically vanish.

Conversely, assume $a(u, v) = b(u, v) = 0$ everywhere. It follows from the affine parameterization (5.5) that X_{uu} is proportional to X_u. Therefore the u-asymptotic tangent line lies entirely on M. Likewise for the v-direction. Hence the affine lift of M is a doubly ruled surface in $\mathbb{R}^3$, and thus a quadric. □

Approximation by quadrics

Consider now an arbitrary non-degenerate surface M. Generically, M can be approximated by a quadric at a point x up to order 2. This follows, for example, from a similar fact in Euclidean geometry.

Proposition 5.1.18 *The points at which M can be approximated by a quadric up to order 3 are the points at which the invariants α and β vanish.*

Proof Choose asymptotic coordinates (u, v) on M and consider an affine lift $X(u, v)$. The linear coordinates (x, y, z) in the affine 3-space can be chosen in such a way that $X(0, 0)$ is the origin and the vectors $X_u(0, 0)$, $X_v(0, 0)$ and

$X_{uv}(0, 0)$ are the unit coordinate vectors. Then the surface is locally given by the equation $z = xy + O(3)$ where $O(3)$ stands for terms cubic in x, y.

Exercise 5.1.19 Check that the equation defining M is

$$z = xy + \frac{1}{3}\left(bx^3 + ay^3\right) + O(4). \tag{5.15}$$

Assume that $a = b = 0$, then the quadric $z = xy$ approximates M up to order 3. Conversely, assume that a quadric approximates M up to order 3. All such quadrics are of the form

$$z = xy + \lambda\, xz + \mu\, yz + \nu\, z^2 \tag{5.16}$$

where λ, μ, ν are arbitrary coefficients. Rewriting this as

$$z = xy + \lambda\, x^2 y + \mu\, xy^2 + O(4)$$

we conclude that the desired approximation implies $a = b = 0$ in (5.15) and, in this case, $\lambda = \mu = 0$ as well. □

Similarly to the notion of an osculating conic for non-degenerate plane curves, one can ask if there is an osculating quadric. The answer is that such quadric is not unique. Formula (5.16) implies the following statement.

Corollary 5.1.20 *At every point of M there is a three-parameter family of quadrics having contact of order 2 with M and, if approximation of order 3 is possible, then there is a one-parameter family of approximating quadrics*

$$z = xy + \nu\, z^2. \tag{5.17}$$

The quadric of Lie

A number of remarkable quadrics having contact of order 2 with a surface is known. One distinguishes a one-parameter family of the Darboux quadrics given by formula (5.17). We will describe only one of the Darboux quadrics called the quadric of Lie.

The geometric definition is as follows. Along with point $x(0, 0)$ of M, consider two points $x(-\varepsilon, 0)$ and $x(\varepsilon, 0)$. Let $\eta, \eta_{-\varepsilon}, \eta_\varepsilon$ be the v-asymptotic tangent lines at these points. These lines determine a quadric Q_ε, made of the lines intersecting the three lines, and the quadric of Lie is the limiting position of Q_ε as $\varepsilon \to 0$.

Exercise 5.1.21 Check that if one interchanges u and v in the above construction, the result is the same quadric.

Hint. Check that the quadric of Lie is given by the equation (5.17) with $\nu = -\frac{1}{2}ab$.

Comment

The first major contribution to the theory of surfaces in $\mathbb{RP}^3$ was made by Wilczynski who introduced the fundamental system of equations and described the differential invariants in five memoirs [232]. The geometric constructions for the differential invariants α and β are due to Bompiani [187]; the constructions for the quadratic form and for the linear projective element are found in [71]. We also recommend the books [129, 29] for the state of the art of projective differential geometry of surfaces in the first half of the twentieth century.

A quadratic surface is one of the oldest objects of mathematics. Second-order contact quadrics were considered already in the nineteenth century. Proposition 5.1.18 is due to Hermite. The quadric of Lie was described in a letter from Lie to Klein dated 1878.

Let us reiterate: our choice of material fits into the general theme of this book. We left aside such aspects as Cartan's *méthode du repère mobile* (see [2, 101, 102] and references therein), projective normals and geodesics, as well as the theory of ruled and developable surfaces, extensively studied in the literature. We also mention an interesting recent relation between system (5.2) and integrable systems; see [63, 64].

5.2 Relative, affine and projective differential geometry of hypersurfaces

In this section we will outline a unified approach to affine and projective differential geometry of hypersurfaces based on the formalism of connections. It is interesting to compare this more recent approach with the more classic material in the previous section. The reader is encouraged to look at Appendix A.3 for relevant basic definitions and results on connections and related topics.

Induced connection on a hypersurface

Much of the differential geometry of a smooth hypersurface M^{n-1} in Euclidean space $\mathbb{R}^n$ can be formulated in terms of the unit normal field. For example, a curve $\gamma(t) \subset M$ is a geodesic if, for all t, the acceleration vector $\gamma''(t)$ belongs to the 2-plane, spanned by the velocity $\gamma'(t)$ and the normal vector at point $\gamma(t)$.

In relative differential geometry, one considers a smooth hypersurface M^{n-1} in n-dimensional affine space V, equipped with a transverse vector field $\nu(x)$. We often consider ν as a vector-valued function on M, so that $d\nu$ is a vector-valued differential 1-form on M. As in the preceding section, we assume throughout this section that M is (quadratically) non-degenerate. We also assume that the ambient affine space V has a fixed volume form Ω, a differential n-form with constant coefficients.

Let $\widetilde{\nabla}$ be a flat affine connection in V: the covariant derivative $\widetilde{\nabla}_v$ is just the directional derivative along the vector v. Let u and v be tangent vector fields on M. Then $\widetilde{\nabla}_v(u)$ can be decomposed into the tangential and transverse components:

$$\widetilde{\nabla}_v(u) = \nabla_v(u) + h(v, u)\nu. \tag{5.18}$$

Exercise 5.2.1 Check that $\nabla_v(u)$ is an affine connection without torsion on M and $h(v, u)$ is a non-degenerate symmetric bilinear form.

The connection $\nabla_v(u)$ is called the *induced connection*. In the classical theory, ν is the unit Euclidean normal, then (5.18) is the Gauss equation and $h(v, u)$ is the second fundamental form. The geodesics of the induced connection are those curves on M whose osculating 2-plane at every point is generated by the velocity vector and the vector ν at this point.

Let v be a tangent vector field on M. One can similarly decompose the covariant derivative of ν:

$$\widetilde{\nabla}_v(\nu) = -S(v) + \theta(v)\nu. \tag{5.19}$$

Here S is a linear transformation of the tangent space TM, called the *shape operator* (the minus sign is a convention), and θ is a 1-form on M. In the classical theory, when ν is the unit Euclidean normal, (5.19) is the Weingarten equation.

Another induced structure on M is the volume form $\omega = i_\nu\Omega$. Its relation with the induced connection is as follows.

Exercise 5.2.2 Let v be a tangent vector field on M. Check that

$$\nabla_v(\omega) = \theta(v)(\omega)$$

where θ is as in (5.19).

Hint. Use the fact that $\widetilde{\nabla}_v(\Omega) = 0$ and the Weingarten equation (5.19).

It is straightforward to compute how a change in the transverse vector field affects the induced connection, the bilinear form h and the 1-form θ.

Exercise 5.2.3 Let $\bar{\nu} = e^{-\psi}\nu + w$ be a new transverse vector field where ψ is a function on M and w is a tangent vector field. Show that

$$\overline{\nabla}_v(u) = \nabla_v(u) - e^{\psi} h(u, v)\, w, \quad \bar{h}(u, v) = e^{\psi} h(u, v), \quad \bar{\theta} = \theta + e^{\psi} i_w h - \mathrm{d}\psi.$$

Relative normalization

So far, the transverse field ν was arbitrary. Now we impose a restriction that the 1-form θ in (5.19) vanishes. A motivation is that this is the case for the Euclidean unit normal. If $\theta = 0$ the transverse field ν is called a *relative normal*. We also call the field of directions generated by a relative normal an *exact transverse line field*; a justification for this term will be provided shortly.

Given a smooth hypersurface $M \subset V$, a conormal at point $x \in M$ is a non-zero covector $p \in T_x^*M$ whose kernel is the tangent hyperplane $T_x M$. A conormal is defined up to a multiplicative constant. If a transverse vector field ν is fixed, we normalize the conormal by the condition $\langle p(x), \nu(x) \rangle = 1$ for all $x \in M$. This defines a conormal covector field $p(x)$ on M that will be considered as a covector-valued function on M. Recall that $\mathrm{d}\nu$ is a vector-valued 1-form on M, therefore $\langle p, \mathrm{d}\nu \rangle$ is a differential 1-form on M.

Lemma 5.2.4 *The transverse vector field ν is a relative normal if and only if $\langle p, \mathrm{d}\nu \rangle = 0$. The latter is also equivalent to $\langle \mathrm{d}p, \nu \rangle = 0$.*

Proof Let v be a tangent vector field on M. Then

$$\langle p, \mathrm{d}\nu \rangle(v) = \langle p, \mathrm{d}\nu(v) \rangle = \langle p, \widetilde{\nabla}_v(\nu) \rangle = \langle p, \theta(v)\nu \rangle = \theta(v)$$

by (5.19). Hence $\theta = 0$ if and only if $\langle p, \mathrm{d}\nu \rangle = 0$.

Finally, $\langle \mathrm{d}p, \nu \rangle = \mathrm{d}\langle p, \nu \rangle - \langle p, \mathrm{d}\nu \rangle$, and the last statement follows. □

The next exercise justifies the term "exact".

Exercise 5.2.5 Let η be a transverse line field along a hypersurface M and let ν be section of η, a transverse vector field. Let p be a conormal field such that $\langle p, \nu \rangle = 1$. Prove that η is exact if and only if the 1-form $\langle p, \mathrm{d}\nu \rangle$ on M is exact. The latter is also equivalent to the exactness of the 1-form $\langle \mathrm{d}p, \nu \rangle$.

It follows from Exercise 5.2.3 that relative normal fields along M are in one-to-one correspondence with non-vanishing functions on M.

Corollary 5.2.6 *Let ν be a relative normal vector field and p the respective conormal field. Then any other relative normal vector field is given by the formula*

$$\bar{\nu} = e^{-\psi}\left(\nu + \text{grad}_h(\psi)\right)$$

where ψ is a smooth function and grad_h *is the gradient with respect to the non-degenerate bilinear form h. The respective conormal field is $\bar{p} = e^{\psi} p$.*

Proof It follows from the last formula in Exercise 5.2.3 that $e^{\psi} i_w h = d\psi$. Therefore $w = e^{-\psi}\text{grad}_h(\psi)$. □

Relative differential geometry of hypersurfaces

Given a relative normalization of a hypersurface M in an affine space, one can develop its differential geometry along the same lines as in the familiar Euclidean case. Since M is non-degenerate, so is the bilinear form h. This form gives M a pseudo-Riemannian metric, sometimes called *relative metric*. Let ∇^h be the respective Levi–Civita connection on M. The following two tensor fields play an important role: the difference $(1, 2)$-tensor

$$K(u, v) = \nabla_v(u) - \nabla_v^h(u) \tag{5.20}$$

and a $(0, 3)$-tensor, the cubic form $C(u, v, w) = \nabla_v(h)(u, w)$.

Exercise 5.2.7 Check that:

(a) $K(u, v) = K(v, u)$;
(b) $C(u, v, w)$ is symmetric in all arguments and

$$C(u, v, w) = -2h(K(u, v), w);$$

(c) $h(S(u), v) = h(u, S(v))$.

The last identity means that the shape operator is self-adjoint with respect to the relative metric. Since h is not necessarily positive-definite, the eigenvalues of S do not have to be all real. However, elementary symmetric functions of these eigenvalues are real, and they play the roles of relative curvatures.

By analogy with the Euclidean case, one introduces the *relative support function*. Fix a point $x_0 \in V$ (an origin) and let x vary on the hypersurface

M. The vector $x - x_0$ can be decomposed into tangential and transverse components:

$$x - x_0 = \rho(x)\,\nu(x) + u, \qquad u \in T_x M. \tag{5.21}$$

The function $\rho : M \to \mathbb{R}$ is the relative support function. If $p(x)$ is the respective conormal then $\rho(x) = \langle p, x - x_0 \rangle$. The tangent vector field u in (5.21) can be recovered from the support function.

Lemma 5.2.8 *One has:* $u = -\text{grad}_h(\rho)$.

Proof Let v be a test tangent vector field on M. One has:

$$\begin{aligned} v = \widetilde{\nabla}_v(x - x_0) &= v(\rho)\,\nu + \rho\widetilde{\nabla}_v(\nu) + \widetilde{\nabla}_v(u) \\ &= v(\rho)\,\nu - \rho S(v) + \nabla_v(u) + h(u, v)\nu. \end{aligned}$$

Equating the transverse components yields $v(\rho) = -h(u, v)$, and it follows that $u = -\text{grad}_h(\rho)$. □

Corollary 5.2.9 *Given a non-degenerate hypersurface M, the relative normal at point $x \in M$ passes through point x_0 if and only if x is a critical point of the respective relative support function.*

Affine differential geometry of hypersurfaces

Let $M \subset V$ be a smooth non-degenerate hypersurface in an affine space. As before, Ω is a volume form in V. The *equiaffine transverse vector field* ν is defined as a relative normalization such that the induced volume form $\omega = i_\nu \Omega$ on M coincides with the volume form, associated with the pseudo-Riemannian metric h. We leave it to the reader to check that this condition defines ν uniquely. The vector $\nu(x)$ is called the affine normal to M at point x. Once the affine normal is defined, affine differential geometry becomes a particular case of relative differential geometry.

The affine normal to a locally convex non-degenerate hypersurface M^{n-1} has the following geometrical interpretation. Let x be a point of M. Consider the one-parameter family of affine hyperplanes, parallel to the tangent hyperplane $T_x M$. Let H be such a hyperplane; its intersection with M bounds an $(n-1)$-dimensional convex domain in H. The locus of centers of mass of these domains is a curve Γ through x. Then the tangent line to Γ at x is the affine normal to M at x; cf. Section 4.1 for the case of curves.

It is not our intention to discuss affine differential geometry in any detail here. Note affine spheres, the hypersurfaces characterized by the property that the affine normals either are all parallel or all pass through one point. If an affine sphere is strictly convex and closed then it is an ellipsoid; however, the situation is much more complicated in the general case. Let us also mention the Pick–Berwald theorem that a non-degenerate hypersurface has a vanishing cubic form if and only if it is a quadric.

Projective invariants of hypersurfaces

Let $M \subset \mathbb{RP}^n$ be a smooth non-degenerate hypersurface. We will construct a powerful projective invariant, a $(1, 2)$-tensor field on M. The reader is referred to Appendix A.3 for information on projective connections.

Choose an equiaffine connection $\widetilde{\nabla}$ and a volume element Ω, representing the canonical (flat) projective connection on $\mathbb{RP}^n$. Consider the respective affine normal vector field to M and let K be the corresponding difference tensor field (5.20) on M.

Theorem 5.2.10 *The* $(1, 2)$*-tensor field* K *does not depend on the choice of the equiaffine connection* $(\widetilde{\nabla}, \Omega)$.

Proof Let $(\widetilde{\nabla}', \Omega')$ be a different choice. Then

$$\Omega' = e^{\varphi}\Omega, \qquad \widetilde{\nabla}'_v(u) = \widetilde{\nabla}_v(u) + \alpha(u)v + \alpha(v)u \tag{5.22}$$

where φ is a function and α a 1-form.

Exercise 5.2.11 Prove that $\alpha = \frac{1}{n+1}\mathrm{d}\varphi$.

Let ν and ν' be the two respective affine normal vector fields, h and h' the corresponding bilinear forms. Then

$$\nu' = e^{\psi}\nu + w \tag{5.23}$$

where ψ is a function and w is a tangent vector field. Combine Gauss equations (5.18) for connections $\widetilde{\nabla}$ and $\widetilde{\nabla}'$ with the second equation (5.22) to obtain

$$\nabla_v(u) + h(u, v)\nu + \alpha(u)\, v + \alpha(v)\, u = \nabla'_v(u) + h'(u, v)\nu'. \tag{5.24}$$

Taking (5.23) into account, equate the transverse, ν-components, to obtain

$$h'(u, v) = e^{-\psi} h(u, v) \tag{5.25}$$

for all tangent vectors u and v.

Next we determine the function ψ. Consider the volume forms $\omega = i_\nu \Omega$ and $\omega' = i_{\nu'} \Omega'$ on M. Let $u_1, \ldots, u_{n-1}$ be local vector fields such that

$$\omega(u_1, \ldots, u_{n-1}) = 1.$$

Since ν is an affine normal, ω coincides with the volume form associated with h, hence $\det h(u_i, u_j) = 1$. It follows from the first equation (5.22) and from (5.23) that $\omega' = e^{\phi+\psi}\omega$. Let

$$u_i' = e^{-(\phi+\psi)/(n-1)} u_i, \quad i = 1, \ldots, n-1,$$

then

$$\omega'(u_1', \ldots, u_{n-1}') = 1.$$

Since ν' is an affine normal, $\det h'(u_i', u_j') = 1$. It now follows from (5.25) that

$$\psi = -\frac{2\phi}{n+1}. \tag{5.26}$$

Next we determine the tangent vector w in (5.23). Since ν' is an affine normal, and in particular a relative normal, the vector $\widetilde{\nabla}_v'(\nu')$ is tangent to M for every tangent vector field v. Using the second equation (5.22) and (5.23), and equating the transverse, ν-components to zero, yields:

$$h(v, w) + e^\psi v(\psi) + e^\psi \alpha(v) = 0.$$

Combining Exercise 5.2.11 and (5.26), one concludes that

$$w = \frac{e^{-2\phi/(n+1)}}{(n+1)} \operatorname{grad}_h(\phi).$$

Now one returns to equation (5.24) and equates the tangential components. One obtains:

$$\nabla_v'(u) = \nabla_v(u) + \frac{1}{n+1}\left(d\varphi(u)\, v + d\varphi(v)\, u - h(u, v) \operatorname{grad}_h(\phi)\right). \tag{5.27}$$

On the other hand, by (5.25) and (5.26),

$$h' = e^{-2\phi/(n+1)} h. \tag{5.28}$$

Exercise 5.2.12 Show that if two pseudo-Riemannian metrics are conformally equivalent as in (5.28) then the corresponding Levi–Civita connections are related as in (5.27).

It follows that $K' = K$, and we are done. □

Surfaces in $\mathbb{RP}^3$ revisited

Let us return to the situation of Section 5.1 and compute the tensor K for a saddle-like surface in $\mathbb{RP}^3$. Let $X(u, v)$ be an affine lift satisfying equations (5.5). The preceding constructions specialize as follows.

Exercise 5.2.13 Check that:

(a) the transverse vector field $e^{-h} X_{uv}$ is a relative normalization;
(b) $\nabla_{X_u} X_u = h_u X_u + b X_v$, $\nabla_{X_u} X_v = \nabla_{X_v} X_u = 0$, $\nabla_{X_v} X_v = a X_u + h_v X_v$;
(c) $K(X_u, X_u) = b X_v$, $K(X_u, X_v) = 0$, $K(X_v, X_v) = a X_u$.

Comment

The "golden period" of affine differential geometry was 1916–1923. The subject was mostly developed by the Blaschke school (Blaschke, Berwald, Pick, Radon, Reidemeister and others); see [26]. Affine differential geometry continued to be an active area of research; let us mention, among many others, the contributions of Calabi and Nomizu that greatly influenced further investigations. Relative differential geometry goes back to work of Müller in the early 1920s. The last 20 years witnessed a renaissance of affine differential geometry and related topics – see [135, 156, 190, 191] for a comprehensive account.

5.3 Geometry of relative normals and exact transverse line fields

In this section we provide examples of relative normals and exact transverse fields along non-degenerate hypersurfaces in affine and projective space and discuss their geometrical properties. We consider relative normals and exact transverse fields as generalizations of normals in Euclidean geometry. The reader is recommended to consult Section A.2 for basic notions of symplectic geometry.

Minkowski normals

Let $M^{n-1} \subset V^n$ be a non-degenerate hypersurface in affine space. We already mentioned that if V has a Euclidean structure then the unit normals ν to M constitute a relative normalization. Indeed, let v be a tangent vector field on M. Then $\langle \nu, \nu \rangle = 1$ and $0 = v(\langle \nu, \nu \rangle) = 2\langle v(\nu), \nu \rangle$. Hence $v(\nu)$ is also tangent to M, and the 1-form θ in (5.19) vanishes.

Consider the following generalization. A *Minkowski metric* in vector space V is given by a smooth, quadratically convex, closed hypersurface S containing the origin. By definition, this hypersurface consists of Minkowski unit vectors. Every vector v is uniquely written as tx where $x \in S$ is unit and $t \geq 0$, and the Minkowski length of v is defined to be equal to t. In other words, a Minkowski metric gives V the structure of a Banach space. Note that we do not assume S to be centrally symmetric. By definition, each vector $x \in S$ is Minkowski orthogonal to the hyperplane $T_x S$. If S is not centrally symmetric then the Minkowski normal line may change if one reverses the coorientation of the hyperplane.

Lemma 5.3.1 *Let $M \subset V$ be a smooth cooriented hypersurface and ν be the field of unit Minkowski normal vectors along M. Then ν is a relative normalization.*

Proof Given a point $x \in S$, let c be the conormal vector to S at x, normalized by $\langle c, x\rangle = 1$. Since c vanishes on $T_x S$, the 1-form $\langle c, dx\rangle$ vanishes on S. Consider the conormal field p along M, normalized by $\langle p, \nu\rangle = 1$. By Lemma 5.2.4, we need to show that $\langle p, d\nu\rangle = 0$ on M. Consider the Gauss map $f : M \to S$ that sends $y \in M$ to $x \in S$ such that $T_y M$ is parallel to $T_x S$ and has the same coorientation. Then $\langle p, d\nu\rangle = f^*(\langle c, dx\rangle) = 0$, and we are done. □

Note the following consequence of the proof. If $M \subset V$ is a smooth hypersurface, star-shaped with respect to the origin, then the transverse vector field $\nu(x) = x$ of position vectors on M is a relative normalization.

Huygens' principle

Consider propagation of light in some medium V. Let F_t be the time-t wave front. For every $x \in F_t$, consider the contact element of F_t at x, that is, the hyperplane $T_x F_t \subset T_x V$, and parallel translate it distance ε in the normal direction to F_t at x. According to Huygens' principle (see [10]), one obtains the family of contact elements to the wave front $F_{t+\varepsilon}$.

A similar property holds for relative normals. Let $M \subset V$ be a smooth hypersurface and ν a relative normal vector field along M. For a fixed positive real t, consider the map $g_t : M \to V$ given by $g_t(x) = x + t\nu(x)$.

Lemma 5.3.2 *The tangent hyperplane to $g_t(M)$ at a smooth point $g_t(x)$ is parallel to $T_x M$.*

Proof Let p be a conormal covector to M at point $x \in M$. Then the 1-form $\langle p, \mathrm{d}x \rangle$ vanishes on M. Let $y = g_t(x)$. Then $T_y(g_t(M))$ is parallel to $T_x M$ if and only if the 1-form $\langle p, \mathrm{d}y \rangle$ vanishes on $g_t(M)$. One has:

$$\langle p, \mathrm{d}y \rangle = \langle p, \mathrm{d}x \rangle + t \langle p, \mathrm{d}\nu \rangle = 0$$

since ν is a relative normal and $\langle p, \mathrm{d}\nu \rangle = 0$ by Lemma 5.2.4. □

Note that, similarly to wave propagation, the "fronts" $g_t(M)$ usually develop singularities; cf. Section 4.6.

Normals in geometries of constant curvature

Let $M \subset V$ be a smooth hypersurface in affine space and assume that V carries a Riemannian metric. In general, it is not true that the normal lines to M constitute an exact line field along M. However, this is true if the geodesics of the Riemannian metric are straight lines. By a Beltrami theorem, this implies that the metric has constant curvature.

More specifically, consider the standard models for geometries of constant curvature ± 1: the unit sphere or the unit pseudosphere. Recall the construction of the latter.

Let H be the upper sheet of the hyperboloid $x^2 - y^2 = -1$ in $\mathbb{R}^n_x \times \mathbb{R}^1_y$ with the Lorentz metric $\mathrm{d}x^2 - \mathrm{d}y^2$. The restriction of the Lorentz metric to H is a metric of constant negative curvature. Project H from the origin to the hyperplane $y = 1$. Let q be the Euclidean coordinate in this hyperplane. The projection is given by the formulae:

$$x = \frac{q}{(1 - |q|^2)^{1/2}}, \quad y = \frac{1}{(1 - |q|^2)^{1/2}},$$

and the image of H is the open unit ball $q^2 < 1$. The hyperbolic metric g in the unit ball is given by the formula:

$$g(u, v) = \frac{\langle u, v \rangle}{1 - |q|^2} + \frac{\langle u, q \rangle \langle v, q \rangle}{(1 - |q|^2)^2} \tag{5.29}$$

where u and v are tangent vectors at q. This is the Klein–Beltrami (or projective) model of hyperbolic space; cf. Section 4.2.

The construction of the metric of constant positive curvature in $\mathbb{R}^n$ is analogous: one replaces the hyperboloid by the unit sphere, and this results in change of signs in formula (5.29).

Proposition 5.3.3 *Let g be a Riemannian metric of constant curvature in a domain $U \subset \mathbb{R}^n$ whose geodesics are straight lines. Then the field of g-normals to any smooth hypersurface is an exact transverse line field.*

Proof We will consider the metric (5.29), the case of positive curvature being completely analogous. Let $H(q, p) : T^*U \to \mathbb{R}$ be the Hamiltonian of the metric g, where p denotes covectors. Lifting the indices in (5.29) yields:

$$H(q, p) = (1 - |q|^2)(|p|^2 - \langle p, q\rangle^2). \tag{5.30}$$

Recall that the geodesic flow in T^*U is the vector field $H_p \partial q - H_q \partial p$, the symplectic gradient of H, and its projection to $T_q U$ is a g-normal to the contact element $p = 0$; see Appendix A.2.

Let M be a hypersurface, q be a point of M and p a conormal at q. We take $\nu = H_p(q, p)/H(q, p)$ as a transverse vector at q. Since H is homogeneous of degree 2 in p, one has $\langle p, \nu\rangle = 2$ by Euler's equation. By Exercise 5.2.5, we need to prove that

$$\frac{1}{H}\langle \mathrm{d}p, H_p\rangle = \langle \mathrm{d}p, (\ln H)_p\rangle$$

is an exact 1-form on M. Indeed, it follows from (5.30) that

$$\langle \mathrm{d}p, (\ln H)_p\rangle = 2\frac{\langle p, \mathrm{d}p\rangle - \langle p, q\rangle\langle q, \mathrm{d}p\rangle}{|p|^2 - \langle p, q\rangle^2} = \mathrm{d}\ln(|p|^2 - \langle p, q\rangle^2),$$

which is exact; the last equality is due to the fact that $\langle p, \mathrm{d}q\rangle$ vanishes on M since p is a conormal. □

The case of spheres

Let η be a smooth transverse line field along S^{n-1}, a round sphere in $\mathbb{R}^n$. Orient the lines from η in the outward direction. Let $\mathcal{L}$ be the space of oriented lines intersecting the sphere. Then η defines an embedding $F_\eta : S^{n-1} \to \mathcal{L}$.

Consider the interior of the sphere as the projective model of hyperbolic space H^n. Then the space of oriented lines $\mathcal{L}$ has a symplectic structure ω, associated with the metric; see Appendix A.2. One has the following characterization of exact line fields. Assume that $n \geq 3$.

Theorem 5.3.4 *The field η is exact if and only if $F_\eta(S^{n-1}) \subset \mathcal{L}$ is a Lagrangian submanifold, that is, $F_\eta^*(\omega) = 0$.*

Proof First, let us calculate the symplectic structure ω. We continue using the notation from the preceding subsection.

To calculate the symplectic form on $\mathcal{L}$ one needs to consider the unit cotangent bundle U^*H^n, restrict the canonical symplectic form Ω from T^*H^n on U^*H^n and factorize by the characteristic foliation; see Appendix A.2. We will construct a certain section $\mathcal{L} \to U^*H^n$ and find ω as the pull-back of Ω.

An oriented line ℓ is characterized by its point $q \in \ell$ and its unit (in the hyperbolic sense) tangent vector u at q. It is convenient to choose q to be the midpoint (in the Euclidean sense) of the segment of ℓ inside the sphere. Thus $\mathcal{L}$ is identified with the following submanifolds of the tangent bundle TH^n:

$$V = \{(q, u) \mid g(u, u) = 1,\ \langle u, q\rangle = 0\}$$

where g is the hyperbolic metric (5.29) and $\langle\ ,\ \rangle$ is the Euclidean scalar product.

Identify the tangent and cotangent bundles by the hyperbolic metric. Then the Liouville 1-form pdq becomes the following form in the tangent bundle

$$\lambda = \frac{\langle u, \mathrm{d}q\rangle}{1 - |q|^2} + \frac{\langle u, q\rangle\langle q, \mathrm{d}q\rangle}{(1 - |q|^2)^2}$$

where u is a tangent vector at point q. Restrict the Liouville 1-form on V. For hyperbolic unit vector u, let v be the proportional Euclidean unit vector. Then $u = (1 - |q|^2)^{1/2} v$ and

$$\lambda|_V = \frac{\langle v, \mathrm{d}q\rangle}{(1 - |q|^2)^{1/2}} = -\frac{\langle \mathrm{d}v, q\rangle}{(1 - |q|^2)^{1/2}}. \tag{5.31}$$

In coordinates (v, q) on $\mathcal{L}$, formula (5.31) describes the symplectic form on $\mathcal{L}$ via $\omega = \mathrm{d}\lambda$.

Now consider the transverse line field η. For $x \in S^{n-1}$, let $v(x)$ be the unit vector along the line $\ell = \eta(x)$. The corresponding midpoint q is easily found: $q = x - \langle x, v\rangle v$, and then (5.31) implies that

$$F_\eta^*(\lambda) = -\frac{\langle \mathrm{d}v, x\rangle}{\langle v, x\rangle}. \tag{5.32}$$

Finally, identify vectors and covectors by the Euclidean structure. Then $p = x/\langle v, x\rangle$ is the conormal to S^{n-1} at x such that $\langle p, v\rangle = 1$. Hence the right hand side of (5.32) is the 1-form $-\langle p, \mathrm{d}v\rangle$. According to Exercise 5.2.5, the line field η is exact if and only if the 1-form $F_\eta^*(\lambda)$ is exact. Since S^{n-1} is simply connected, this is equivalent to $F_\eta^*(\omega) = 0$, and the result follows. □

Remark 5.3.5 For $n = 2$, the formulation of Theorem 5.3.4 needs to be adjusted: the curve F_η, which is automatically Lagrangian, should be exact Lagrangian.

Applying Lemma A.2.5, one obtains the next corollary.

Corollary 5.3.6 *The field η is exact if and only if the lines from η are the hyperbolic normals to a closed hypersurface in H^n, diffeomorphic to the sphere. In fact, there is a one-parameter family of equidistant hypersurfaces, perpendicular to the lines from η.*

The next characterization of exact transverse line fields along spheres will be useful in the next section.

Exercise 5.3.7 Let f be a smooth function on the unit sphere. Using Corollary 5.2.6, show that the transverse vector field $\nu(x) = x + \operatorname{grad} f(x)$ generates an exact line field, and every exact transverse line field is of this form for a suitable function f.

Relative normals to plane curves

Consider a closed immersed parameterized plane curve $\gamma(t)$ and let $\eta(t)$ be a transverse line field along γ. We denote by $\times$ the cross-product of vectors in the plane. Let $\nu(t)$ be a section of $\eta(t)$, a transverse vector field generating the line field η.

Lemma 5.3.8 *The line field η is exact if and only if*

$$\int_\gamma \frac{\nu \times \gamma_{tt}}{\gamma_t \times \nu}\, \mathrm{d}t = 0.$$

Proof Let us use the cross-product to identify vectors and covectors. Then $p = \gamma_t$ is a conormal field along γ. Consider another transverse vector field $\nu_1 = \nu/(\gamma_t \times \nu)$; then $\langle p, \nu_1 \rangle = 1$. According to Lemma 5.2.4, η is exact if and only if the 1-form $\eta_1\, \mathrm{d}p$ is exact on γ. The latter 1-form equals $(\nu \times \gamma_{tt})/(\gamma_t \times \nu)\mathrm{d}t$, and the result follows. □

Exercise 5.3.9 Show that the integral in Lemma 5.3.8 does not depend on the parameterization of the curve.

Let γ be a non-degenerate curve so that γ_t and γ_{tt} are linearly independent. It follows from Lemma 5.3.8 that the acceleration vectors γ_{tt} generate an exact transverse line field. The converse holds as well.

Lemma 5.3.10 *Given an exact transverse line field η along a non-degenerate curve γ, there is a parameterization $\gamma(\tau)$ such that η is generated by the vector field $\gamma_{\tau\tau}$.*

Proof Let $\gamma(t)$ be some parameterization, and let ν be a section of η. According to Lemma 5.3.8, the 1-form $(\nu \times \gamma_{tt})/(\gamma_t \times \nu)\, dt$ is exact, that is,

$$\nu \times \gamma_{tt} = f_t\, (\gamma_t \times \nu) \tag{5.33}$$

for some function f on the curve. We are looking for a new parameter, $\tau(t)$, such that ν is collinear with $\gamma_{\tau\tau}$. By (5.33) and since the velocity and acceleration vectors are everywhere linearly independent, this is equivalent to

$$\gamma_{\tau\tau} \times \gamma_{tt} = f_t\, (\gamma_t \times \gamma_{\tau\tau}). \tag{5.34}$$

By the Chain Rule, one has:

$$\gamma_\tau = \gamma_t\, t_\tau, \quad \gamma_{\tau\tau} = \gamma_{tt} t_\tau^2 + \gamma_t t_{\tau\tau},$$

and (5.34) is rewritten as $t_{\tau\tau} = f_t t_\tau^2$ or $t_{\tau\tau} = f_\tau t_\tau$. The latter equation is easily solved: $(\ln t_\tau)_\tau = f_\tau$ and hence $t_\tau = ce^f$ where c is a constant. This gives the desired parameterization. □

Recall that vertices of a convex plane curve correspond to singular points of the envelop of the family of its normals; see Section 4.1. One has the following version of the four-vertex theorem for exact transverse line fields.

Theorem 5.3.11 *Let γ be a convex closed plane curve, η a generic exact transverse line field along γ and Γ the envelop of the family of lines from η. Assume that the lines from η revolve in the same sense as one traverses γ. Then Γ has at least four cusps.*

Proof Every function on a circle has a critical point. By Corollary 5.2.9, for every point of the plane there is a line from η passing through this point.

The tangent lines to Γ are the lines from η. Since the lines from η revolve in the same sense, infinitesimally close lines are not parallel. Hence Γ is compact. Since the tangent direction to Γ revolves in the same sense all the time, Γ is also locally convex, and its total turning angle is 2π. That is, the Gauss map of Γ is one-to-one. If Γ has no cusps then it is a closed convex curve, and there are no tangent lines to Γ from points inside it. This contradicts the previous paragraph. Since Γ has a coorientation induced by an orientation of γ, the number of cusps of Γ is even. Thus Γ has at least two cusps.

Suppose Γ has only two cusps. Consider a locally constant function $\phi(x)$ in the complement of Γ whose value at point x equals the number of tangent lines to Γ through x. The value of this function increases by 2 as x crosses Γ from locally concave to locally convex side; see Figure 5.5. Let x be sufficiently far away from Γ. Since the Gauss map is one-to-one, $\phi(x) = 2$.

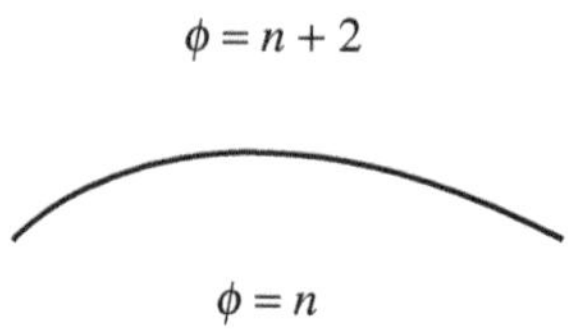

Fig. 5.5. Number of tangent lines

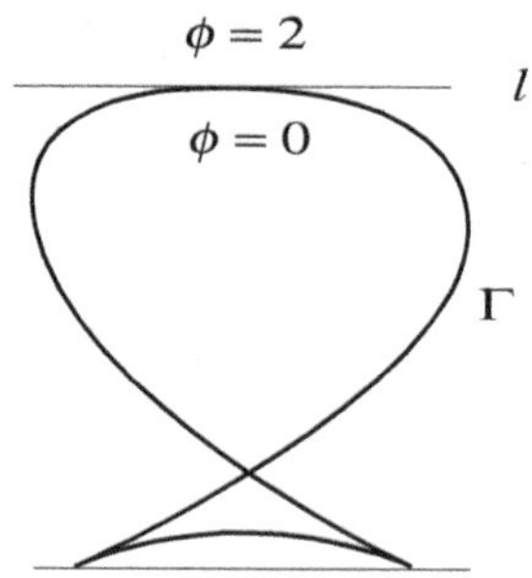

Fig. 5.6. Case of two cusps

Consider the line through two cusps of Γ (which may coincide); assume it is horizontal; see Figure 5.6. Then the height function, restricted to Γ, attains either minimum or maximum (or both) not in a cusp. Assume it is maximum; draw the horizontal line l through it. Since Γ lies below this line, $\phi = 2$ above it. Therefore $\phi(x) = 0$ immediately below l, and there are no tangent lines to Γ from x. This again contradicts the first paragraph of the proof. □

Projective invariance

The notion of exact transverse line field is defined in affine terms; however, exactness is projectively-invariant.

Let $M^{n-1} \subset \mathbb{RP}^n$ be a smooth hypersurface and η a transverse line field along M. Consider the projection $\pi : \mathbb{R}^{n+1} - 0 \to \mathbb{RP}^n$ and denote by $\widetilde{M}$ the preimage of M. Then $\widetilde{M}$ is a conical manifold, invariant under homotheties. Lift the field η to a transverse line field $\widetilde{\eta}$ along $\widetilde{M}$. We say that η is *exact in the projective sense* if $\widetilde{\eta}$ is an exact line field along $\widetilde{M}$.

Proposition 5.3.12 *This definition of exactness does not depend on the choice of the lift $\widetilde{\eta}$. If M lies in an affine chart, this definition is equivalent to the one given in Section 5.2.*

Proof Denote by E the Euler vector field in $\mathbb{R}^{n+1}$. This field generates the fibers of π. As a function $\mathbb{R}^{n+1} \to \mathbb{R}^{n+1}$, the field E is given by the formula $E(x) = x$.

Assume that $\widetilde{\eta}$ is exact. Let $\widetilde{\nu}$ be a section of $\widetilde{\eta}$ and $\widetilde{p}$ be a conormal field along $\widetilde{M}$ such that $\langle \widetilde{p}, \widetilde{\nu} \rangle = 1$ and $\langle \mathrm{d}\widetilde{p}, \widetilde{\nu} \rangle = 0$. Let $\widetilde{\eta}_1$ be a different lift of η, generated by the vector field $\widetilde{\nu}_1 = \widetilde{\nu} + fE$ where f is a function. Since E is tangent to $\widetilde{M}$, one has $\langle \widetilde{p}, E \rangle = 0$ and therefore $\langle \widetilde{p}, \widetilde{\nu}_1 \rangle = 1$. Also $\langle \mathrm{d}\widetilde{p}, E \rangle + \langle \widetilde{p}, \mathrm{d}E \rangle = 0$. But $\langle \widetilde{p}, \mathrm{d}E \rangle = \langle \widetilde{p}, \mathrm{d}x \rangle = 0$ on $\widetilde{M}$ since $\widetilde{p}$ is a conormal. It follows that $\langle \mathrm{d}\widetilde{p}, E \rangle = 0$, and hence $\langle \mathrm{d}\widetilde{p}, \widetilde{\nu}_1 \rangle = 0$. Thus $\widetilde{\nu}_1$ is exact.

To prove the second statement, choose coordinates $(x_0, x_1, \ldots, x_n)$ in $\mathbb{R}^{n+1}$ and identify the affine part of $\mathbb{RP}^n$ with the hyperplane $x_0 = 1$. Decompose $\mathbb{R}^{n+1}$ into $\mathbb{R}^1_{x_0} \oplus \mathbb{R}^n_x$ where $x = (x_1, \ldots, x_n)$; vectors and covectors are decomposed accordingly.

Assume that M belongs to the hyperplane $x_0 = 1$. Let ν be a relative normalization and η the exact line field generated by ν. We want to show that η is exact in the projective sense, that is, a lifted line field $\widetilde{\eta}$ is exact.

Let p be the conormal field such that $\langle p, \nu \rangle = 1$ and $\langle \mathrm{d}p, \nu \rangle = 0$. Consider the covector field

$$\widetilde{p}(x_0, x) = \left(-\left\langle \frac{x}{x_0}, p\left(\frac{x}{x_0}\right) \right\rangle, p\left(\frac{x}{x_0}\right) \right)$$

along $\widetilde{M}$. Then $\langle \widetilde{p}, E \rangle = 0$, hence $\widetilde{p}$ is a conormal field on $\widetilde{M}$. Lift ν to the horizontal vector field $\widetilde{\nu}(x_0, x) = (0, \nu(x/x_0))$. Then $\langle \widetilde{p}, \widetilde{\nu} \rangle = 1$ and $\langle \mathrm{d}\widetilde{p}, \widetilde{\nu} \rangle = 0$. Hence η is exact in the projective sense.

Conversely, if $\widetilde{\nu}(x_0, x) = (0, \nu(x/x_0))$ is a relative normalization of $\widetilde{M}$ then ν is a relative normalization of M. □

As a consequence of Proposition 5.3.12, exactness of a transverse line field is a projective property. More precisely, if $M \subset V$ is a smooth hypersurface in an affine space, η is an exact transverse field along M and $f : V \to V$ is a projective transformation whose domain contains M then the line field $\mathrm{d}f(\eta)$ along $f(M)$ is exact as well.

Exercise 5.3.13 Let $M \subset V$ be a smooth hypersurface in an affine space, ν a relative normalization, p a conormal field satisfying $\langle p, \nu \rangle = 1$ and $\langle \mathrm{d}p, \nu \rangle = 0$. Let $\ell(x)$ be a linear function on V. Consider the projective map $f : V \to V$ given by the formula $f(x) = x/(1 + \ell(x))$. This map takes M to a hypersurface $\overline{M}$, the vector field ν to a vector field $\bar{\nu}$ and the covector field p

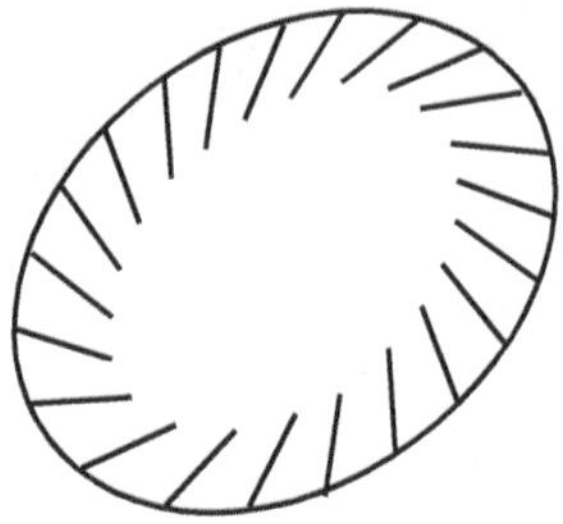

Fig. 5.7. Non-exact transverse line field

to a conormal field $\bar{p}$ along $\overline{M}$. Show that

$$\langle \mathrm{d}\bar{p}, \bar{\nu} \rangle = \mathrm{d} \ln(1 + \ell(x)).$$

Number of relative normals through a point

Let M be a non-degenerate closed immersed hypersurface with a relative normalization ν in an affine space V, and let $x_0 \in V$ be a fixed point. According to Corollary 5.2.9, the number of relative normals passing through x_0 is not less than the number of critical points of a smooth function on M. In particular, there are at least two relative normals passing through x_0. This Morse theory type result is a generalization of a similar fact for Euclidean normals. Without the exactness assumption, this results does not hold; see Figure 5.7.

Consider a more general situation: M is a non-degenerate closed immersed hypersurface in $\mathbb{RP}^n$ equipped with a transverse line field η, exact in the projective sense. An example of a non-degenerate surface is a hyperboloid in $\mathbb{RP}^3$ or its sufficiently small perturbation.

Theorem 5.3.14 *The number of lines from η, passing through a fixed point $x_0 \in \mathbb{RP}^n$, is bounded below by the least number of critical points of a smooth function on M.*

For example, for the hyperboloid in $\mathbb{RP}^3$, this number is 3, and it equals 4 for a generic choice of x_0.

Proof We use the same notation as in the preceding subsection: $\widetilde{M} \subset \mathbb{R}^{n+1}$ is a conical hypersurface, $\widetilde{\nu}$ and $\widetilde{p}$ are its relative normalization and a conormal field satisfying $\langle \widetilde{p}, \widetilde{\nu} \rangle = 1$ and $\langle \mathrm{d}\widetilde{p}, \widetilde{\nu} \rangle = 0$. We choose $\widetilde{\nu}$ and $\widetilde{p}$ to be homogeneous of degrees 1 and -1, respectively. Let $y_0 \in \mathbb{R}^{n+1}$ be a point that projects to x_0.

Consider the support function $\widetilde{\rho}(x) = \langle \widetilde{p}, x - y_0 \rangle$. Let $\widetilde{h}$ be the bilinear form on $\widetilde{M}$ associated with its relative normalization. Since $\widetilde{M}$ is a conical manifold, the form $\widetilde{h}$ is no longer non-degenerate: it has a kernel, generated by the Euler field E. Formula (5.21) still holds:

$$x - y_0 = \widetilde{\rho}(x)\widetilde{\nu}(x) + u, \qquad u \in T_x\widetilde{M},$$

and the proof of Lemma 5.2.8 yields that $v(\widetilde{\rho}) = -\widetilde{h}(u, v)$ for every vector v, tangent to $\widetilde{M}$ at point x. If x is a critical point of $\widetilde{\rho}$ then u belongs to the kernel of $\widetilde{h}$, that is, u is proportional to E. Hence the vector $x - y_0$ projects to the line $\eta(\pi(x))$, which therefore passes through $x_0 = \pi(y_0)$.

By our choice of homogeneity, the support function $\widetilde{\rho}$ is homogeneous of degree zero. Therefore this function descends to M, and we are done. □

Comment

The results of this section are contained in [203, 206, 207, 208]. Tabachnikov [208] contains a definition of exact line fields in terms of cross-ratio, similar to the geometric constructions in Section 5.1. Another result of [208] is an extension of the notion of exactness to line fields along convex polyhedral hypersurfaces; in particular, there is an analog of Theorem 5.3.11 for polygons; see [210]. One should not expect the analogy between exact line fields and Euclidean normals to go too far. For example, a smooth convex closed hypersurface $M \subset \mathbb{R}^n$ has at least n diameters (i.e., double normals) but there exists an exact line field η along M such that, for all pairs of distinct points $x, y \in M$, one has: $\eta(x) \neq \eta(y)$; see [208].

5.4 Complete integrability of the geodesic flow on the ellipsoid and of the billiard map inside the ellipsoid

In this section we apply results on exact transverse line fields along spheres from the preceding section to prove complete integrability of two closely related classical dynamical systems: the geodesic flow on the ellipsoid and the billiard map inside an ellipsoid. This complete integrability has been known since the first half of the nineteenth century. Again we refer to Appendix A.2 for necessary notions of symplectic geometry.

Geodesic flow on the ellipsoid and billiard map inside it

The geodesic flow on a Riemannian manifold M^n is a flow on the tangent bundle TM: a tangent vector v moves with constant speed along the geodesic, tangent to

v. From the physical viewpoint, the geodesic flow describes the motion of a free particle on M. Identifying the tangent and cotangent bundles by the metric, the geodesic flow becomes a Hamiltonian vector field on the cotangent bundle T^*M with its canonical symplectic structure "$dp \wedge dq$", the Hamiltonian function being the energy $H(q, p) = |p|^2/2$. Complete integrability of the geodesic flow means that the flow has n invariant functions (integrals), independent on an open dense set and Poisson commuting with respect to the canonical symplectic structure (cf. Appendix A.2).

Let M be a compact convex domain with a smooth boundary in $\mathbb{R}^{n+1}$. The billiard system describes the motion of a free particle inside M with elastic reflections off the boundary. One replaces the continuous time system by its discrete time reduction, the billiard transformation. The billiard transformation T is a transformation of the set of oriented lines in $\mathbb{R}^{n+1}$ that intersect M; the map T is defined by the familiar law of geometric optics: the incoming ray ℓ, the outgoing ray $T(\ell)$ and the normal to the boundary ∂M at the impact point lie in one 2-plane, and the angles made by ℓ and $T(\ell)$ with the normal are equal.

The space $\mathcal{L}$ of oriented lines in Euclidean space $\mathbb{R}^{n+1}$ is a $2n$-dimensional symplectic manifold, and the billiard transformation is a symplectomorphism; see, for example, [198]. Complete integrability of T means that there exist n integrals, functionally independent on an open dense subset of $\mathcal{L}$ and Poisson commuting with respect to this symplectic structure ω.

Let an ellipsoid $E \subset \mathbb{R}^{n+1}$ be given by the equation

$$\sum_{i=1}^{n+1} \frac{x_i^2}{a_i^2} = 1, \tag{5.35}$$

and assume that all the semiaxes a_i are distinct. The family of confocal quadratic hypersurfaces E_t is given by the equation

$$\sum_{i=1}^{n+1} \frac{x_i^2}{a_i^2 + t} = 1, \quad t \in \mathbb{R},\ t \neq -a_i^2.$$

If E is an ellipse (i.e., $n = 1$) then E_t consists of ellipses and hyperbolas with the same foci as E.

The geometric meaning of complete integrability is as follows. Consider a fixed geodesic γ on E. Then the straight lines tangent to γ are also tangent to $n - 1$ fixed confocal hypersurfaces $E_{t_1}, \ldots, E_{t_{n-1}}$, where the parameters $t_1, \ldots, t_{n-1}$ depend on γ. Thus one has $n - 1$ integrals of the geodesic flow, and one more integral is the energy. Likewise, consider an oriented line ℓ intersecting E. Then ℓ is tangent to n confocal hypersurfaces, and all the reflected lines $T(\ell)$, $T^2(\ell)$, ... remain tangent to the same n confocal hypersurfaces.

Exercise 5.4.1 Let E be an ellipse. Show the following. If an oriented line ℓ does not pass between the foci of E then ℓ and all the reflected lines are tangent to the same confocal ellipse; if ℓ passes between the foci then ℓ and all the reflected lines remain tangent to the same confocal hyperbola; and if ℓ passes through a focus then the reflected line passes through another focus.

The billiard transformation inside the ellipsoid and the geodesic flow on the ellipsoid are closely related. On the one hand, the domain inside the ellipsoid (5.35) can be considered as a degenerate ellipsoid, the limit of the ellipsoids

$$\sum_{i=1}^{n+2} \frac{x_i^2}{a_i^2} = 1$$

as $a_{n+2} \to 0$. Thus the billiard system is obtained from the geodesic flow; in particular, the integrals of the latter are those of the former.

On the other hand, consider a free particle inside the ellipsoid E whose trajectory meets E at a small angle α. In the limit $\alpha \to 0$, one obtains a free particle on E, and the integrals of the billiard transformation inside E provide integrals of the geodesic flow on E.

Bisymplectic maps and projectively equivalent metrics

Consider the following simple mechanism providing integrals of the billiard transformation. Let M^{2n} be a smooth manifold with two symplectic forms ω, Ω and let T be a diffeomorphism that preserves both. We will call M a *bisymplectic manifold* and T a *bisymplectomorphism*. Then the functions f_i defined by the relation

$$\Omega^i \wedge \omega^{n-i} = f_i \, \omega^n, \quad i = 1, \dots, n \tag{5.36}$$

are T-invariant. If the forms ω and Ω are sufficiently generic then these integrals are functionally independent almost everywhere. The same construction works for a degenerate 2-form Ω.

Alternatively, a bisymplectic structure determines a field E of linear automorphisms of tangent spaces:

$$\Omega(u, v) = \omega(Eu, v) \quad \text{for all } u, v \in T_x M. \tag{5.37}$$

The eigenvalues of E all have multiplicity 2; they are invariant functions, and these n integrals functionally depend on the integrals $f_1, \dots, f_n$.

Remark 5.4.2 In general, the integrals $f_1, \ldots, f_n$ are not in involution with respect to either of the symplectic structures. If ω and Ω are Poisson compatible (the sum of the respective Poisson structures is again a Poisson structure) then the integrals $f_1, \ldots, f_n$ Poisson commute with respect to both forms – see, for example, [140] concerning bihamiltonian formalism. However Poisson compatibility is not necessary for the functions $f_1, \ldots, f_n$ to be in involution.

Now we modify the above integrability mechanism to provide integrals of the geodesic flow. Consider a smooth manifold M^n with two Riemannian metrics g_1, g_2. We call the metrics *projectively (or geodesically) equivalent* if their non-parameterized geodesics coincide.

Theorem 5.4.3 *If a manifold M^n carries two projectively equivalent metrics then the geodesic flow of each has n integrals.*

Proof Consider a Riemannian manifold (M^n, g_1). The cotangent bundle T^*M has a canonical symplectic form, and this form is the differential of the Liouville 1-form. Identifying T^*M with TM by the metrics g_1, one obtains a symplectic form ω_1 and 1-form λ_1 on the tangent bundle such that $\omega_1 = \mathrm{d}\lambda_1$. Let $S_1 \subset TM$ be the unit vector hypersurface. The form λ_1 is a contact form on S_1, that is, $\lambda_1 \wedge \omega_1^{n-1}$ is a non-degenerate volume form. The restriction of ω_1 to S_1 has a one-dimensional kernel ξ at every point, and the curves tangent to ξ are the trajectories of the geodesic flow of the metric g_1; cf. Appendix A.2. The same applies to the second metric g_2.

Consider the map $\phi : TM \to TM$ that rescales the tangent vectors, sending S_1 to S_2. This map sends the trajectories of the geodesic flow of g_1 to those of g_2. Consider the two forms $\omega = \omega_1$ and $\Omega = \phi^*(\omega_2)$. These forms have the same characteristic foliation on the hypersurface S_1 and both are holonomy invariant along this foliation. It follows that the functions f_i, defined by the equality

$$\lambda_1 \wedge \omega^{n-1-i} \wedge \Omega^i = f_i \, \lambda_1 \wedge \omega^{n-1}, \quad i = 1, \ldots, n-1, \tag{5.38}$$

are integrals of the geodesic flow of the metric g_1, the n-th integral being the energy. These integrals are analogs of the integrals (5.36).

Alternatively, the forms ω and Ω determine non-degenerate 2-forms on the quotient $2(n-1)$-dimensional spaces TS_1/ξ. As before, one may consider the field of automorphisms E, analogous to (5.37), given by the formula:

$$\Omega(u, v) = \omega(Eu, v) \quad \text{for all } u, v \in T_xM/\xi.$$

The eigenvalues of E have multiplicities 2, and they are invariant along the characteristic foliation. □

Projective billiards

The billiard dynamical system is defined in metric terms (equal angles). Let us define a broader class of "billiards" with a projectively invariant law of reflection, called *projective billiards*. Let M be a smooth hypersurface in projective space and η a smooth field of transverse directions along M. The law of the projective billiard reflection reads: the incoming ray, the outgoing ray and the line $\eta(x)$ at the impact point x lie in one 2-plane π, and these three lines, along with the line of intersection of π with the tangent hyperplane $T_x M$, constitute a harmonic quadruple of lines.

Recall that four coplanar and concurrent lines constitute a harmonic quadruple if the cross-ratio of these lines equals -1. The cross-ratio of four coplanar concurrent lines is the cross-ratio of their intersection points with a fifth line (it does not depend on the choice of this auxiliary line).

If M lies in Euclidean space and η consists of Euclidean normals to M then the projective billiard coincides with the usual one. More generally, consider a metric g in a domain U in a vector space whose geodesics are straight lines; this is a metric of constant curvature. Let M be a smooth hypersurface and η the field its g-normals.

Lemma 5.4.4 *The billiard reflection in M associated with the metric g is a projective billiard transformation.*

Proof Let $x \in M$ be the impact point. The metric g gives the tangent space $T_x U$ a Euclidean structure, and therefore the billiard reflection in $T_x M \subset T_x U$ is a projective one. The projective structures in U and $T_x U$ coincide, and the result follows. □

As a limit case of the above lemma, one obtains a similar result for the geodesic flow.

Lemma 5.4.5 *Let γ be a g-geodesic line on M. Then, for every $x \in \gamma$, the osculating 2-plane to γ at x contains the line $\eta(x)$. In other words, γ is a geodesic of the connection ∇, associated with the field of g-normals η; see Section 5.2.*

Integrability

We are in a position to establish complete integrability of the geodesic flow on the ellipsoid and the billiard map inside the ellipsoid. Let us start with the latter.

Let η be an exact transverse line field along the sphere $S^n \subset \mathbb{R}^{n+1}$. Consider the interior of the sphere as the Klein–Beltrami model of hyperbolic space H^{n+1}, and denote by Ω the corresponding symplectic structure on the space of oriented lines in H^{n+1}; cf. Section 5.3. Let T be the projective billiard map associated with the line field η.

Proposition 5.4.6 *One has:* $T^*(\Omega) = \Omega$.

Proof According to Corollary 5.3.6, there exists a one-parameter family of equidistant hypersurfaces that are perpendicular, in the hyperbolic sense, to the lines from the family η. Choose one of these hypersurfaces, say, M_0, and denote the hypersurfaces in the family by M_t where t is the hyperbolic distance from M_0 to M_t along the normals. Consider the billiard transformation inside M_t; it preserves the symplectic structure Ω. According to Lemma 5.4.4, this billiard map is the projective billiard transformation inside M_t associated with the field η, which therefore preserves Ω. As $t \to \infty$, the limit of the billiard transformations inside M_t is the projective billiard map T inside S^n, and the result follows. □

Proof of complete integrability of the billiard map inside the ellipsoid. Let $E \subset \mathbb{R}^{n+1}$ be an ellipsoid and η_1 the field of Euclidean normals along E; this field is exact. Apply an affine transformation ψ that takes E to the unit sphere S^n, and let $\eta = \psi(\eta_1)$. Then η is an exact field of lines. The map ψ conjugates the original billiard transformation inside E and the projective billiard transformation T inside S^n associated with η. According to Proposition 5.4.6, Ω is a T-invariant 2-form on the space of lines $\mathcal{L}$.

On the other hand, the original billiard transformation inside E preserves the canonical symplectic structure on the space of oriented lines associated with the Euclidean metric in $\mathbb{R}^{n+1}$, and therefore T also preserves a symplectic structure ω on $\mathcal{L}$ associated with the Euclidean metric induced by the affine map ψ. Thus T is a bisymplectic map, and it remains to feed ω and Ω to the integrability "mechanism" of Theorem 5.4.3. One needs to check that the resulting integrals Poisson commute; the can be done by a direct computation which we omit.

Let us now consider the geodesic flow on the ellipsoid. One has the following analog of Proposition 5.4.6.

Proposition 5.4.7 *Let η be an exact transverse line field along S^n and ∇ the corresponding connection on the sphere. Then there exists a Riemannian metric g on the sphere whose non-parameterized geodesics coincide with the geodesics of ∇.*

Proof As in the proof of Proposition 5.4.6, consider the family of equidistant hypersurfaces M_t, orthogonal to the lines η. Let h_t be the metric on M_t induced from the ambient hyperbolic space. By Lemma 5.4.5, this metric has the same geodesics as the connection associated with η. The intersections with the lines from the family η determine an identification of each M_t with S^n, and we consider h_t as metrics on the sphere. The metrics h_t exponentially grow with t but they have a limit after a renormalization. Namely, the desired metric on the sphere is given by the formula

$$g = \lim_{t\to\infty} e^{-t} h_t, \tag{5.39}$$

and this completes the proof. □

Exercise 5.4.8 Let the field η be generated by the transverse vector field $x + \operatorname{grad} f(x)$; see Exercise 5.3.7. Show that the metric (5.39) is conformally equivalent to the Euclidean metric on S^n with the conformal factor e^{-f}.

Everything is prepared for the final touch.

Proof of complete integrability of the geodesic flow on the ellipsoid. Arguing as before, let $E \subset \mathbb{R}^{n+1}$ be an ellipsoid, η_1 the field of Euclidean normals, ψ an affine transformation that takes E to S^n and $\eta = \psi(\eta_1)$. Let g be the metric (5.39). Then the induced metric $\psi^*(g)$ on E is projectively equivalent to the Euclidean one, and it remains to apply Theorem 5.4.3. Again it is a separate step to check that the integrals Poisson commute; we do not dwell on this.

An explicit formula for the second metric is given in the following theorem that summarizes the argument; we leave it as an exercise to deduce this result from Exercise 5.4.8.

Theorem 5.4.9 *The restrictions of the metrics*

$$\sum_{i=1}^{n+1} \mathrm{d}x_i^2 \qquad \text{and} \qquad \frac{\sum_{i=1}^{n+1} a_i\, \mathrm{d}x_i^2}{\sum_{i=1}^{n+1} a_i^2\, x_i^2}$$

on the ellipsoid $\sum_{i=1}^{n+1} a_i\, x_i^2 = 1$ *are projectively equivalent.*

Remark 5.4.10 The standing assumption that the ellipsoid has distinct semi-axes a_i is essential for obtaining the right number of integrals for complete integrability. For example, if all a_i are equal then the statement of Theorem 5.4.9 becomes tautological. However, the integrals provided by the above construction

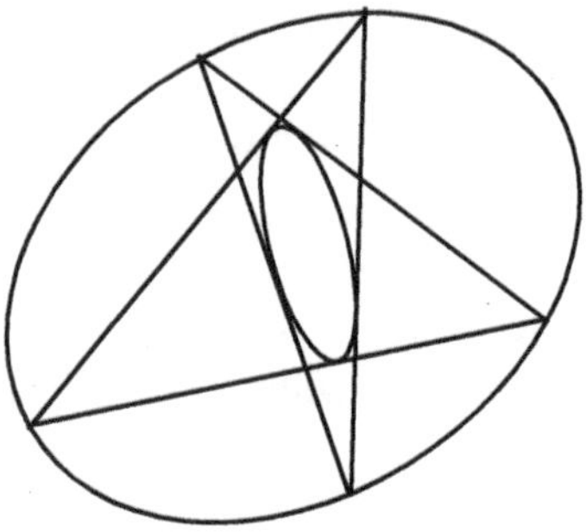

Fig. 5.8. Poncelet configuration

"survive" in the non-generic case and remain functionally independent almost everywhere.

Comment

The complete integrability of the geodesic flow on the triaxial ellipsoid was established by Jacobi in 1838. Jacobi integrated the geodesic flow by separation of variables. The appropriate coordinates are called the elliptic coordinates, and this approach works in any dimension. Two other proofs of the complete integrability of the geodesic flow on the ellipsoid, by confocal quadrics and by isospectral deformations, are described in [150, 151, 152].

G. Birkhoff was the first to put forward the study of mathematical billiards; the complete integrability of the billiard inside the ellipsoid is discussed in [23]. See [153, 198, 226] for contemporary proofs and [227] for complete integrability of the billiard inside the ellipsoid in a space of constant curvature.

The integrability of the billiard transformation inside an ellipse implies the famous Poncelet porism. Given two nested ellipses A, B in the plane, one plays the following game: choose a point $x \in B$, draw a tangent line to A through it, find the intersection y with B, and iterate, taking y as a new starting point. The statement is that if x returns back after a number of iterations, then every point of B will return back after the same number of iterations; see Figure 5.8. See [22, 30] for the interesting history and various proofs of this theorem.

The proof of integrability in this section follows the papers [203, 209, 212]; projective billiards were introduced in [205]. Our proof of complete integrability applies without change to the billiard map inside and the geodesic flow on the ellipsoid in the spherical or the hyperbolic space. Theorems 5.4.3 and 5.4.9 were independently discovered by Matveev and Topalov [144, 145, 146, 147]. In particular, they proved that, for projectively equivalent Riemannian metrics, the integrals f_i in (5.38) Poisson commute.

5.5 Hilbert's fourth problem

We saw in the previous section that the existence of sufficiently general Riemannian metrics, projectively equivalent to a given one, implies the complete integrability of its geodesic flow. In this section we discuss the simplest integrable case, the Euclidean metric in a convex domain in $\mathbb{R}^n$. The problem is to describe all Finsler metrics, projectively equivalent to the Euclidean one.

Formulation of the problem and examples

In his fourth problem, Hilbert proposes to "construct and study the geometries in which the straight line segment is the shortest connection between two points". Hilbert was motivated by two interesting examples, well understood by the time he delivered his celebrated 1900 ICM lecture. The first of these examples is Minkowski geometry which we briefly discussed in Section 5.3.

The second example was discovered by Hilbert himself in 1894, and it is called the *Hilbert metric*. Hilbert's metric generalizes the Klein–Beltrami model of hyperbolic geometry. Consider a convex closed hypersurface $S \subset \mathbb{R}^n$ and define the distance between points inside S by formula (4.5). If S is a sphere or, more generally, an ellipsoid, one has the Klein–Beltrami model of hyperbolic geometry.

Exercise 5.5.1 Prove that the Hilbert distance satisfies the triangle inequality.

As we mentioned above, a Riemannian metric projectively equivalent to a Euclidean one is a metric of constant curvature (Beltrami's theorem) and therefore either a Euclidean, or a spherical, or a hyperbolic metric. Hence it does not make sense to restrict Hilbert's fourth problem to Riemannian metrics only. The adequate class consists of *Finsler metrics*, briefly introduced below (see, e.g, [17, 180] for details).

Finsler metric

Finsler geometry describes the propagation of light in an inhomogeneous anisotropic medium. This means that the velocity depends on the point and the direction. There are two equivalent descriptions of this process corresponding to the Lagrangian and the Hamiltonian approaches in classical mechanics; we will mostly consider the former.

A Finsler metric on a smooth manifold M is described by a smooth field of strictly convex smooth hypersurfaces containing the origin in the tangent space at each point. We also assume that the hypersurfaces are centrally-symmetric (this assumption is sometimes omitted). These hypersurfaces are called *indicatrices*. The indicatrix consists of the Finsler unit vectors and plays the role of the unit sphere in Riemannian geometry which is a particular case of Finsler geometry.

Equivalently, a Finsler metric is determined by a non-negative fiber-wise convex Lagrangian function $L(x, v)$ where $x \in M,\ v \in T_x M$, on the tangent bundle TM whose unit level hypersurface S intersects each fiber of TM along the indicatrix; the function L is smooth off the zero section $v = 0$. We make a standing assumption that the Lagrangian is fiber-wise homogeneous of degree 1:

$$L(x, tv) = |t|\, L(x, v), \quad t \in \mathbb{R}. \tag{5.40}$$

Thus L gives each tangent space a Banach norm. For the usual Euclidean metric, $L(x, v) = |v|$.

Given a smooth curve $\gamma : [a, b] \to M$, its length is given by

$$\mathcal{P}(\gamma) = \int_a^b L(\gamma(t), \gamma'(t))dt.$$

As a result of (5.40), the integral does not depend on the parameterization. A Finsler geodesic is an extremal of the functional $\mathcal{P}$. The Finsler geodesic flow is a flow in TM in which the foot point of a vector in TM moves along the Finsler geodesic tangent to it, so that the vector remains tangent to this geodesic and preserves its norm.

Let $I_x \subset T_x M$ be the indicatrix and $u \in I_x$. Let $p \in T_x^* M$ be the conormal to I_x, normalized by $\langle p, u \rangle = 1$. The mapping $u \mapsto p$ is called the *Legendre transform*, its image is a smooth strictly convex hypersurface in the cotangent space. This hypersurface is called the *figuratrix*, and it is the unit sphere of the dual normed space $T_x^* M$. In local coordinates, $p = L_u$. If the Finsler metric is Riemannian then the Legendre transform is the identification of the tangent and cotangent spaces by the metric. The figuratrix is the unit level surface of a Hamiltonian function $H : T^*M \to \mathbb{R}$, homogeneous of degree 1 in the momentum p. The Hamiltonian flow of this function on T^*M is also called the Finsler geodesic flow. The Legendre transform identifies the two flows; cf. Appendix A.2.

Hamel's theorem

We interpret Hilbert's fourth problem as asking to describe Finsler metrics in convex domains in $\mathbb{R}^n$ whose geodesics are straight lines. Such Finsler metrics are called *projective*. Hamel's theorem of 1901 is the first general result on projective metrics (it also holds for non-reversible metrics for which $L(x, -v) \neq L(x, v)$). The formulation makes use of a local coordinate system.

Theorem 5.5.2 *A Lagrangian $L(x, v)$ defines a projective Finsler metric if and only if the matrix of second partial derivatives $L_{xv}(x, v)$ is symmetric for all (x, v).*

Proof Assume that the extremals of L are straight lines. Let $x(t)$ be an extremal, $u = x'$. The Euler–Lagrange equation for extremals reads:

$$L_{uu}x'' + L_{ux}x' - L_x = 0. \tag{5.41}$$

Since the extremals are straight lines, x'' is proportional to u. Since L_u is homogeneous of degree 1, the Euler's equation implies: $L_u u = L$ and, differentiating, $L_{uu}u = 0$. Hence, by (5.41), $L_{ux}u = L_x$. In coordinates,

$$\sum_k L_{u_i x_k} u_k = L_{x_i}$$

for all i. Differentiate with respect to u_j to obtain:

$$\sum_k L_{u_i u_j x_k} u_k = L_{x_i u_j} - L_{x_j u_i}$$

for all i, j. The left hand side is symmetric in i, j while the right hand side is skew-symmetric. Therefore both vanish, and $L_{x_i u_j} = L_{x_j u_i}$.

Conversely, let the matrix $L_{x_i u_j}$ be symmetric. Then $L_{ux}u = L_{xu}u = L_x$, and $L_{ux}u - L_x = 0$. The Euler–Lagrange equation (5.41) implies that $L_{uu}x'' = 0$. The matrix $L_{uu}(x, u)$ is degenerate and its kernel is generated by the vector u. Thus x'' is proportional to x', and the extremals are straight lines. □

Exercise 5.5.3 Check that the Lagrangian

$$L(x_1, x_2, v_1, v_2) = \frac{1}{\sqrt{v_1^2 + v_2^2}}\left((3 + x_1^2 + x_2^2)(v_1^2 + v_2^2) + (x_1 v_1 + x_2 v_2)^2\right)$$

defines a projective Finsler metric in the plane (this example is borrowed from [3]).

In a certain sense, Hamel's theorem solves Hilbert's fourth problem. However this theorem gives no clue how to obtain the Lagrangians satisfying its conditions. For example, the Lagrangian of Exercise 5.5.3 looks rather mysterious.

Solution in dimension 2

Let us describe the solution of Hilbert's fourth problem in dimension 2. Our exposition follows [3].

A synthetic approach, due to Busemann, makes use of integral geometry, namely, the Crofton formula (see, e.g., [181]).

Consider the set $\mathcal{L}$ of oriented lines in the plane; topologically, $\mathcal{L}$ is the cylinder. Let an oriented line $\ell \in \mathcal{L}$ have the direction β. One characterizes ℓ by the angle $\alpha = \beta - \pi/2$ and the signed distance p from ℓ to the origin. The 2-form $\omega_0 = \mathrm{d}p \wedge \mathrm{d}\alpha$ is an area form on $\mathcal{L}$; this symplectic form is a particular case of a symplectic structure on the space of oriented geodesics; see Appendix A.2, especially Exercise A.2.4. The Crofton formula expresses the Euclidean length of a plane, not necessarily closed, curve γ in terms of ω_0. The curve determines a locally constant function $N_\gamma(\ell)$ on $\mathcal{L}$, the number of intersections of a line ℓ with γ.

Theorem 5.5.4 *One has:*

$$\text{length}\,(\gamma) = \frac{1}{4}\int_{\mathcal{L}} N_\gamma(\ell)\,|\omega_0| \tag{5.42}$$

where $|\omega_0|$ is the measure associated with the symplectic form ω_0.

Exercise 5.5.5 Prove the Crofton formula (5.42).

Hint. Assume that γ is a polygonal line. Since both sides of (5.42) are additive, it suffices to consider the case when γ is a segment, which is done by a direct computation.

Let $f(p, \alpha)$ be a positive continuous function on $\mathcal{L}$, even with respect to the orientation reversion of a line: $f(-p, \alpha + \pi) = f(p, \alpha)$. Then

$$\omega = f(p, \alpha)\,\mathrm{d}p \wedge \mathrm{d}\alpha \tag{5.43}$$

is also an area form on the space of oriented lines.

Lemma 5.5.6 *Formula (5.42), with ω replacing ω_0, defines a projective Finsler metric.*

Proof To prove that the geodesics are straight lines one needs to check the triangle inequality: the sum of lengths of two sides of a triangle is greater than the length of the third side. This holds because every line intersecting the third side also intersects the first or the second. □

To prove that this integral-geometric construction provides all projective Finsler metrics and to have an explicit formula for the Lagrangians, consider an analytic approach to the problem, due to Pogorelov [176].

Theorem 5.5.7 *The Lagrangians satisfying Hamel's condition of Theorem 5.5.2 have the following integral representation:*

$$L(x_1, x_2, v_1, v_2) = \int_0^{2\pi} |v_1 \cos\phi + v_2 \sin\phi| \; g(x_1 \cos\phi + x_2 \sin\phi, \phi) d\phi$$

where g is a smooth positive function, even in ϕ.

For example, the metric of Exercise 5.5.3 corresponds to $g(p, \alpha) = 1 + p^2$.

Proof Fix x, then L becomes a function of the velocity only. In polar coordinates in $T_x\mathbb{R}^2$, one has: $L(r, \alpha) = rp(\alpha)$. Let I be the indicatrix at point x; its equation is $r = 1/p(\alpha)$. Consider $p(\alpha)$ as the support function of a curve J; cf. Section 4.1.

Exercise 5.5.8 Prove that the curve J is the figuratrix at point x.

Parameterize J by the angle ϕ, made by its tangent vector with the horizontal axis, and let $\rho(\phi)$ be the radius of curvature at point $J(\phi)$. One has:

$$J'(\phi) = \rho(x, \phi)(\cos\phi, \sin\phi),$$

and hence

$$J(\alpha + \pi/2) = J(0) + \int_0^{\alpha+\pi/2} \rho(\phi)(\cos\phi, \sin\phi) d\phi.$$

Write a similar equation for $J(\alpha + 3\pi/2)$, subtract from $J(\alpha + \pi/2)$ and take scalar product with the vector $(\cos\alpha, \sin\alpha)$ to obtain

$$p(\alpha) = \frac{1}{2} \int_{\alpha-\pi/2}^{\alpha+\pi/2} \cos(\alpha - \phi)\rho(\phi) d\phi,$$

and hence

$$L(x_1, x_2, r, \alpha) = \frac{r}{2} \int_{\alpha-\pi/2}^{\alpha+\pi/2} \cos(\alpha - \phi)\rho(x_1, x_2, \phi) d\phi. \tag{5.44}$$

Next, rewriting differential operators $\partial^2/\partial x_i \partial v_j$ in polar coordinates and using (5.44), one expresses the Hamel condition of Theorem 5.5.2 as

$$\int_{\alpha-\pi/2}^{\alpha+\pi/2} \left(-\sin\phi \frac{\partial \rho}{\partial x_1} + \cos\phi \frac{\partial \rho}{\partial x_2} \right) \mathrm{d}\phi = 0$$

for all α. It follows that the integrand vanishes identically, and therefore

$$\rho(x_1, x_2, \phi) = f(x_1 \cos\phi + x_2 \sin\phi, \phi) \tag{5.45}$$

for some function f. Setting $g = f/4$ and rewriting (5.44) in Cartesian coordinates, one obtains the desired result. □

It remains to connect Lemma 5.5.6 and Theorem 5.5.7, the integral-geometric and analytic constructions. In one direction, the connection is as follows.

Exercise 5.5.9 Show that f in (5.43) is the same as f in (5.45). More precisely, the Lagrangian of the metric of Lemma 5.5.6, corresponding to a function $f(p, \alpha)$, is given by Theorem 5.5.7 with $g(p, \alpha) = f(p, \alpha)/4$.

Hint. Apply the construction of Lemma 5.5.6 to an infinitesimally small segment and repeat the computation from Exercise 5.5.5.

The converse relation between the two constructions is as follows. Consider a Finsler metric of Theorem 5.5.7. This metric provides the space of oriented lines with a symplectic structure; see Appendix A.2.

Exercise 5.5.10 In the notation of the proof of Theorem 5.5.7, show that this symplectic structure is given by formula (5.43).

Multi-dimensional case

Let us outline the solution of Hilbert's fourth problem in higher dimensions.

The integral-geometric approach of Busemann and Pogorelov is still applicable. Consider the space $\mathcal{H} = \mathbb{R} \times S^{n-1}$ of oriented affine hyperplanes in $\mathbb{R}^n$ and a signed measure $\rho(h)\mathrm{d}h$ on it. Given a curve γ, one defines a locally constant function $N_\gamma(h)$ on $\mathcal{H}$, the number of intersections of a hyperplane h with γ. Then one defines the ρ-length of γ by the Crofton formula

$$\text{length}\,(\gamma) = \int_{\mathcal{H}} N_\gamma(h)\, \rho(h)\mathrm{d}h. \tag{5.46}$$

This length corresponds to a Lagrangian $L(x, v)$, homogeneous of degree 1, so that

$$\text{length}\,(\gamma) = \int L(\gamma(t), \gamma'(t))\mathrm{d}t.$$

The extremals of this Lagrangian are straight lines (analog of Lemma 5.5.6), and every such Lagrangian is given by formula (5.46) with an appropriate ρ. The latter statement is proved using the next integral representations, generalizing that of Theorem 5.5.7.

Theorem 5.5.11 *A Lagrangian $L(x, v)$, homogeneous of degree 1, satisfies Hamel's condition of Theorem 5.5.2 if and only if there exists a smooth even function $\nu(p, \xi)$ on $\mathbb{R} \times S^{n-1}$ such that*

$$L(x, v) = \int_{\xi \in S^{n-1}} |\langle v, \xi \rangle|\, \nu(\langle v, \xi \rangle, \xi)\mathrm{d}\mu$$

where μ is the standard volume form on the unit sphere $S^{n-1} \subset \mathbb{R}^n$.

To guarantee that formula (5.46) indeed gives Finsler metrics one needs to impose the following positivity conditions on the measure ρ: given three non-collinear points x, y, z, the measure of the set of hyperplanes intersecting twice the wedge xy and yz is positive.

Another approach to the multi-dimensional Hilbert's fourth problem is by way of special symplectic structures on the space $\mathcal{L}$ of oriented lines in $\mathbb{R}^n$; this approach is due to Alvarez. Note that for $n = 2$ the two approaches merge since a line is also a hyperplane.

Let ω be a symplectic form on $\mathcal{L}$. Given two points x and y in $\mathbb{R}^n$, choose a 2-plane P containing both, and let $\mathcal{L}(P)$ be the set of lines lying in P. The restriction of ω on $\mathcal{L}(P)$ is a closed 2-form therein, and one may define the length of the segment xy by the Crofton formula, as in Lemma 5.5.6. It is clear that one obtains a projective metric in P.

However, one wants to make sure that this length does not depend on the choice of the plane P. A sufficient condition is that ω is *admissible*. A symplectic form ω on $\mathcal{L}$ is called admissible if, for every point $x \in \mathbb{R}^n$, the submanifold of lines through x is Lagrangian in $\mathcal{L}$, and if ω is odd under the involution of $\mathcal{L}$ changing the orientation of a line.

One can prove that every projective Finsler metric in $\mathbb{R}^n$ corresponds to an admissible symplectic structure on the space of oriented lines; see [5, 4].

Comment

Two classic references on Hilbert's fourth problem are [36, 176]; see also [197]. The problem is an example of an inverse problem of the calculus of variations (see, e.g., [50]): one wants to describe all variational problems whose (non-parameterized) extremals are given. A magnetic version of Hilbert's fourth problem in the plane is solved in [213]; the problem is to describe the Finsler metrics whose geodesics are circles of a given radius.

5.6 Global results on surfaces

In this section we survey, without proofs, some global results and conjectures on surfaces in projective spaces, old and new.

Carathéodory's conjecture on umbilic points

Let M be a smooth surface in $\mathbb{R}^3$. Recall from classical differential geometry that, at every point of M, two principal curvatures are defined. A point $x \in M$ is called *umbilic* if the two principal curvatures are equal. In other words, a sphere (or a plane) is tangent to M at x with order 2. Also recall that if x is not an umbilic point then one has two orthogonal *principal directions* at x: the orthogonal sections of M in these directions have extremal curvatures. The integral curves of the field of principal directions are called *lines of curvature*; umbilic points are the singularities of this field.

In the early 1920s, in search of a version of the four-vertex theorem in dimension 3, Carathéodory conjectured that a sufficiently smooth convex closed surface in $\mathbb{R}^3$ has at least two distinct umbilic points. This conjecture has a long, interesting and somewhat controversial history.

The only known approach to Carathéodory's conjecture is local. Let $x \in M$ be an isolated umbilic point. Each of the two fields of principal directions has a singularity at x, and this singularity has an index. Since a principal direction is a line without a preferred orientation, this index is a half-integer; see Figure 5.9 for the case of a general ovaloid. By the Poincaré–Hopf theorem, the sum of indices over all umbilic points is equal to the Euler characteristic of M, that is, 2, and the conjecture would follow if one could prove that the index of an isolated singularity of this foliation cannot exceed 1. The last statement is what numerous mathematicians working on the problem over the years aimed at. This statement is stronger than Carathéodory's conjecture and makes the convexity assumption irrelevant.

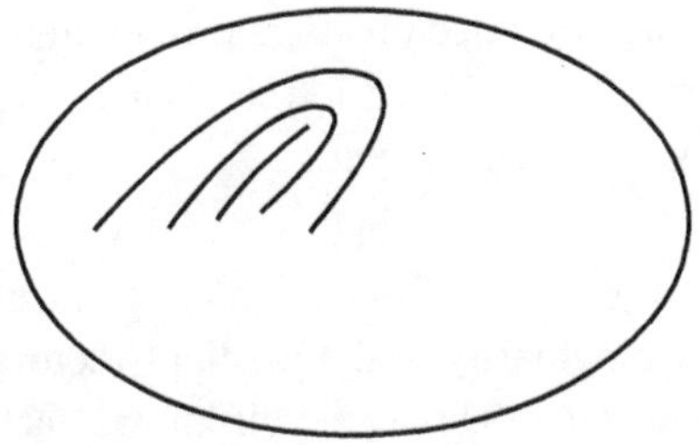

Fig. 5.9. Index of an umbilic point

In special, Ribaucour, coordinates near an umbilic point, the field of principal directions is given by the equation

$$f_{xy}(\mathrm{d}x^2 - \mathrm{d}y^2) + (f_{yy} - f_{xx})\mathrm{d}x\mathrm{d}y = 0 \tag{5.47}$$

where $f(x, y)$ is a function, determined by the germ of the surface, analytic for a real analytic surface.

Lines of curvature on surfaces in $\mathbb{R}^3$ is an old subject, Monge studied them in the late eighteenth century. Approximately 100 years later, Darboux described local geometry of lines of curvature; see [47]. The first proof of Carathéodory's conjecture for real analytic surfaces appeared in a series of long and complicated papers by Hamburger [93, 94, 95], and it was soon simplified by Bol [27]. These proofs did not totally satisfy the mathematical community, and 15 years later Klotz published her proof [119], correcting shortcomings in the previous works. This pattern continued over the years; let us mention a paper by Titus [221] and, as far as we know, the most recent attempt, by Ivanov [100]. To quote from this 72 page long paper: "First, considering analytic surfaces, we assert with full responsibility that Carathéodory was right. Second, we know how this can be proved rigorously. Third, we intend to exhibit here a proof which, in our opinion, will convince every reader who is really ready to undertake a long a tiring journey with us."

Note that a certain degree of smoothness is needed for Carathéodory's conjecture to hold; see [19] for a counterexample.

Loewner's conjecture

About 1950, Ch. Loewner formulated a conjecture whose particular case implies that of Carathéodory. Let f be a real analytic function on the open unit disc. Assume that the vector field $\partial^n f/\partial \bar{z}^n$ has an isolated zero at the origin. Then the index of this zero does not exceed n. The case of $n = 2$ is relevant to umbilics; cf. (5.47).

The status of Loewner's conjecture is approximately the same as that of Carathéodory's. The proof in [221] is not free of shortcomings, and a complete understandable proof is not available yet.

Remark 5.6.1 The $n = 1$ case of Loewner's conjecture is the statement that the index of an isolated singularity of the gradient vector field of a function of two variables cannot exceed 1. This is a well-known fact, proved by standard techniques of differential equations. Interestingly, this fact was used to prove one of the first results of emerging symplectic topology: an orientation and area preserving diffeomorphism of S^2 has at least two distinct fixed points (see, e.g., [15, 148]). One cannot help thinking that Carathéodory's and Loewner's conjectures are similarly related to a yet unknown version of multi-dimensional Sturm theory.

Two reformulations of Carathéodory's conjecture

Let us discuss two reformulations of Carathéodory's conjecture that may shed new light on this fascinating problem.

First, consider the support function $p : S^2 \to \mathbb{R}$ of the convex surface M; cf. Section 4.1. Let x, y, z be Cartesian coordinates in $\mathbb{R}^3$. The support functions of spheres are the restrictions on the unit sphere S^2 of affine functions $ax + by + cz + d$. A version of Carathéodory's conjecture reads therefore as follows:

Given a smooth function p on S^2, there are at least two distinct points at which the 2-jet of p coincides with that of an affine function $ax + by + cz + d$.

This formulation resembles the four-vertex theorem which states that if $p : S^1 \to \mathbb{R}$ is the support function of a closed plane curve then there exist four distinct points at which the 3-jet of p coincides with that of an affine function $ax + by + c$; see Section 4.1.

Next, consider a smooth cooriented surface $M \subset \mathbb{R}^3$. Assign to a point $x \in M$ the oriented normal line to M at x. One obtains an immersion ψ of M to the space $\mathcal{L}$ of oriented lines in $\mathbb{R}^3$. The image of this immersion is Lagrangian with respect to the symplectic structure on $\mathcal{L}$; see Appendix A.2.

Consider an almost complex structure on $\mathcal{L}$ defined as follows. Given a line $\ell \in \mathcal{L}$, consider the rotation of space about ℓ through $\pi/2$. The differential of this rotation is a linear transformation J of the tangent space $T_\ell \mathcal{L}$ whose fourth iteration is the identity, that is, an almost complex structure.

Exercise 5.6.2 Prove that J is a complex structure.

Hint. Use the diffeomorphism $\mathcal{L} = T^*S^2$ and the complex structure on S^2.

In terms of J, the umbilics are described as follows.

Lemma 5.6.3 *A point $x \in M$ is umbilic if and only if $\psi(x)$ is a complex point of $\psi(M)$.*

Proof Consider the quadratic surface that is tangent to M at x with order 2. The point x is umbilic if and only if this quadric is invariant under the rotation through $\pi/2$ about the normal line $\ell = \psi(x)$. The latter is equivalent to the tangent space $T_\ell\psi(M) \subset T_\ell\mathcal{L}$ being invariant under the linear transformation J. □

Thus a version of Carathéodory's conjecture can be stated as follows:

A Lagrangian immersion $S^2 \to \mathcal{L}$ has at least two distinct complex points.

It is interesting what the least number of umbilics is on a surface of genus 2 or higher; a torus can be embedded in space without umbilic points at all. The fact that the index of a singularity of the field of principal directions cannot exceed 1 does not help answering this question.

Arnold's conjectures on non-degenerate hypersurfaces

A classic Hadamard theorem asserts that a closed positively curved hypersurface in Euclidean space bounds a convex domain (see e.g., [193]). Consider a non-degenerate smooth cooriented hypersurface $M^{n-1} \subset \mathbb{RP}^n$, $n \geq 3$. The signature (k, l), $k + l = n - 1$, of the second fundamental form of M is projectively well defined and remains the same at every point of M. A generalization of the Hadamard theorem states that if M has signature $(n - 1, 0)$ then M lies in an affine chart and bounds a convex domain therein; see [8].

In [8], V. Arnold proposed a number of conjectures generalizing this result to non-degenerate hypersurfaces M of signature (k, l):

(1) *One of the components of the complement to M contains a space $\mathbb{RP}^k$ and the other a space $\mathbb{RP}^l$.*
(2) *Every projective line, connecting a point of $\mathbb{RP}^k$ with a point of $\mathbb{RP}^l$, transversally intersects M at two points.*
(3) *M is diffeomorphic to the quotient space of $S^k \times S^l$ by the antipodal involution $(x, y) \mapsto (-x, -y)$.*

(4) *The space of embeddings of a hypersurface of signature* (k, l) *is connected.*

The main example is a quadratic hypersurface, given in $\mathbb{RP}^n$ by the equations $x_1^2 + \ldots + x_{k+1}^2 = y_1^2 + \ldots + y_{l+1}^2$, and its small perturbations.

Arnold's conjectures remain mostly open, even for saddle-like surfaces in $\mathbb{RP}^3$. In a recent remarkable series of papers [108, 109, 110], Khovanskii and Novikov obtain a number of results around these conjectures; in particular, they prove an affine version of Conjecture (1). This version requires an assumption on the behavior of the hypersurface at infinity. One says that a hypersurface M approaches a hypersurface N at infinity if M and N are arbitrarily C^2-close outside a sufficiently large ball. One of the results of Khovanskii and Novikov is as follows.

Theorem 5.6.4 *Let* $M \subset \mathbb{R}^n$ *be a non-degenerate hypersurface of signature* (k, l), $k + l = n - 1$, *which approaches the quadratic cone* $x_1^2 + \ldots + x_{k+1}^2 = y_1^2 + \ldots + y_l^2$ *at infinity. Then one of the components of the complement to* M *contains a* k*-dimensional affine subspace and the other an* l*-dimensional one.*

A slightly different asymptotic behavior may change the situation drastically. For example, there exists a surface in $\mathbb{R}^3$ which approaches the surface $x^2 + y^2 = (|z| - 1)^2$ at infinity and whose complement contains no lines.

6

Projective structures on smooth manifolds

In this chapter we consider M, a smooth manifold of dimension n. How does one develop projective differential geometry on M? If M is a $\mathrm{PGL}(n+1, \mathbb{R})$-homogeneous space, locally diffeomorphic to $\mathbb{RP}^n$, then the situation is clear, but the supply of such manifolds is very limited. Informally speaking, a projective structure on M is a local identification of M with $\mathbb{RP}^n$ (without the requirement that the group $\mathrm{PGL}(n+1, \mathbb{R})$ acts on M). Projective structure is an example of the classic notion of a G-structure widely discussed in the literature. Our aim is to study specific properties of projective structures; see [121, 220] for a more general theory of G-structures.

There are many interesting examples of manifolds that carry projective structures; however, the general problem of existence and classification of projective structures on an n-dimensional manifold is wide open for $n \geq 3$. There is a conjecture that every three-dimensional manifold can be equipped with a projective structure; see [196]. This is a very hard problem and its positive solution would imply, in particular, the Poincaré conjecture.

In this chapter we give a number of equivalent definitions of projective structures and discuss some of their main properties. We introduce two invariant differential operators acting on tensor densities on a manifold and give a description of projective structures in terms of these operators. We also discuss the relation between projective structures and contact geometry. In the two-dimensional case, we present a classification of projective structures.

6.1 Definitions, examples and main properties

In this section we define the notion of projective structure on a smooth manifold and discuss its basic properties.

Definitions

Let us start with three equivalent definitions of projective structure on an n-dimensional manifold M.

(a) An atlas (U_i, x_i), where (U_i) is a countable covering of M by open sets and $x_i = (x_i^1, \ldots, x_i^n) : U_i \to \mathbb{R}^n$ are coordinate maps, is called a *projective atlas* if the transition functions $x_j \circ x_i^{-1}$ are *linear-fractional*, that is, in $U_i \cap U_j$, one has

$$x_i^k = \frac{a_0^k + a_1^k x_j^1 + \cdots + a_n^k x_j^n}{a_0^0 + a_1^0 x_j^1 + \cdots + a_n^0 x_j^n} \tag{6.1}$$

where, for each (i, j), the $(n+1) \times (n+1)$-matrix $A = (a_\ell^k)$ is non-degenerate (and defined up to a multiplicative constant). Two projective atlases are called equivalent if their union is again a projective atlas. A class of equivalent projective atlases is called a *projective structure*.

A projective structure is called an *affine structure* if the transition functions are given by affine transformations, more precisely, $a_1^0 = \cdots = a_n^0 = 0$ in (6.1).

(b) Consider an atlas (U_i, φ_i) where the maps $\varphi_i : U_i \to \mathbb{RP}^n$ are such that the maps $\varphi_i \circ \varphi_j^{-1}$ are projective, that is, are given by the restriction of an element of $\mathrm{PGL}(n+1, \mathbb{R})$ to $\varphi_j(U_j) \subset \mathbb{RP}^n$. Equivalence of such atlases is defined in the same way as in (a), and an equivalence class is called a projective structure.

In the case of affine structures, the transition functions $\varphi_i \circ \varphi_j^{-1}$ are given by the restriction of elements of $\mathrm{Aff}(n, \mathbb{R}) \subset \mathrm{PGL}(n+1, \mathbb{R})$ where $\mathrm{Aff}(n, \mathbb{R})$ is the *affine subgroup* of $\mathrm{PGL}(n+1, \mathbb{R})$, that is, the subgroup preserving an arbitrary fixed hyperplane $\mathbb{RP}^{n-1} \subset \mathbb{RP}^n$.

(c) Let $\widetilde{M}$ be the universal covering of M. A smooth map

$$\varphi : \widetilde{M} \to \mathbb{RP}^n \tag{6.2}$$

is called a *developing map* if it is a local diffeomorphism and if there is a homomorphism

$$T : \pi_1(M) \to \mathrm{PGL}(n+1, \mathbb{R}) \tag{6.3}$$

such that for all $\gamma \in \pi_1(M)$ and $x \in \widetilde{M}$ one has

$$\varphi(\gamma(x)) = T_\gamma \circ \varphi(x); \tag{6.4}$$

here we consider the fundamental group as a group of deck transformations of $\widetilde{M}$. A projective structure on M is given by a developing map modulo equivalence: $\varphi \sim A\varphi$ for some

$A \in \mathrm{PGL}(n+1, \mathbb{R})$. The homomorphism (6.3) is called the *monodromy* (or holonomy) of the projective structure.

A projective structure is affine if there is a hyperplane $\mathbb{RP}^{n-1} \subset \mathbb{RP}^n$ which does not intersect the image $\varphi(\widetilde{M})$ and the range of the homomorphism (6.3) is $\mathrm{Aff}(n, \mathbb{R})$.

Exercise 6.1.1 Prove that the definitions (a) – (c) are equivalent.

sl($n+1$, $\mathbb{R}$), the symmetry algebra of projective structure

The Lie group $\mathrm{PGL}(n+1, \mathbb{R})$ is the group of symmetries in projective geometry. The corresponding Lie algebra is $\mathrm{sl}(n+1, \mathbb{R})$.

Exercise 6.1.2 The standard $\mathrm{sl}(n+1, \mathbb{R})$-action on $\mathbb{RP}^n$ is generated by the vector fields

$$\frac{\partial}{\partial x^i}, \qquad x^j \frac{\partial}{\partial x^i}, \qquad x^j \sum_{i=1}^{n} x^i \frac{\partial}{\partial x^i} \tag{6.5}$$

where $(x^1, \ldots, x^n)$ are affine coordinates. Show that the vector fields in (6.5) with constant and with linear coefficients generate the standard affine subalgebra $\mathrm{aff}(n, \mathbb{R}) \subset \mathrm{sl}(n+1, \mathbb{R})$.

It follows that the space of vector fields generated by (6.5) is stable with respect to the linear-fractional coordinate transformations (6.1). One can prove that, conversely, a projective atlas is such an atlas that the (n^2+2n)-dimensional space of vector fields spanned by the fields (6.5) is stable with respect to the transition functions.

Remark 6.1.3 Note that the $\mathrm{sl}(n+1, \mathbb{R})$-action on M is defined only locally; once a developing map is fixed, one can define an action of $\mathrm{sl}(n+1, \mathbb{R})$ globally on $\widetilde{M}$.

Examples of projective structures

Let us start with simple general constructions producing a substantial supply of examples.

First, if M_1 and M_2 are two manifolds of the same dimension carrying projective structures then their connected sum $M_1 \# M_2$ can be given a projective structure. Namely, one removes a disc from each manifold and identifies collars of the boundary spheres by a projective transformation.

Second, if Γ is a discrete subgroup of $\mathrm{SL}(n+1,\mathbb{R})$, acting properly on S^n, then the quotient S^n/Γ has a natural projective structure. The most famous of these examples, after $\mathbb{RP}^n$ itself, is the Poincaré sphere; see, for example, [111] for an account of various facets of this space.

Example 6.1.4 Every closed surface of genus $g \geq 2$ has a projective structure. Indeed, the hyperbolic plane H^2 has a projective structure since the Klein–Beltrami model realizes H^2 as the interior of a disc in $\mathbb{RP}^2$; see Section 4.2. The surface is the quotient space of H^2 by a discrete group of isometries which acts by projective transformations.

Example 6.1.5 Every projective structure on S^n with $n \geq 2$ is diffeomorphic to the standard one. Indeed, consider a developing map $\varphi : S^n \to \mathbb{RP}^n$. Since S^n is simply connected, this map lifts to $\widetilde{\varphi} : S^n \to S^n$. The immersion $\widetilde{\varphi}$ is a diffeomorphism. Therefore the initial projective structure on S^n is the pull-back of the standard one.

Interesting explicit examples can be constructed in the two-dimensional case.

Example 6.1.6 (Sullivan–Thurston). Consider a (generic) element A of $\mathrm{PGL}(3,\mathbb{R})$, represented by a matrix

$$A = \begin{pmatrix} e^a & 0 & 0 \\ 0 & e^b & 0 \\ 0 & 0 & e^c \end{pmatrix}$$

with $a > b > c$. The matrix A corresponds to the $t = 1$ map for the flow A_t on $\mathbb{RP}^2$, given in homogeneous coordinates by

$$(x : y : z) \mapsto (xe^{at} : ye^{bt} : ze^{ct}).$$

The flow A_t has three fixed points; see Figure 6.1. Consider a closed curve $\gamma \in \mathbb{RP}^2$, transverse to the flow lines. Such a curve is characterized by a word in two symbols; see Figure 6.1.

Two curves γ and $A\gamma$ bound an immersed annulus, and the projective map A identifies the boundary components. Gluing these components according to the identification, one obtains a torus $\mathbb{T}^2$ with projective structure induced from $\mathbb{RP}^2$. The monodromy of this projective structure associates the operator A to one of the generators of $\pi_1(\mathbb{T}^2) = \mathbb{Z}^2$ and the identity to the second one.

Example 6.1.7 Let M be a two-dimensional *open* manifold, for instance, a compact surface with $n \geq 1$ discs removed; then M can be endowed with an

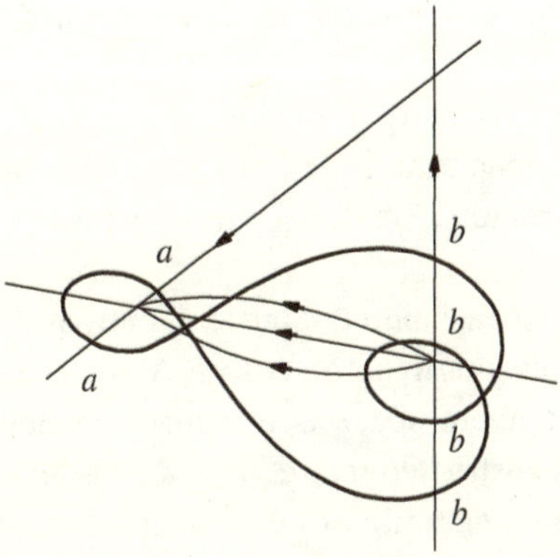

Fig. 6.1. *aabbbb*

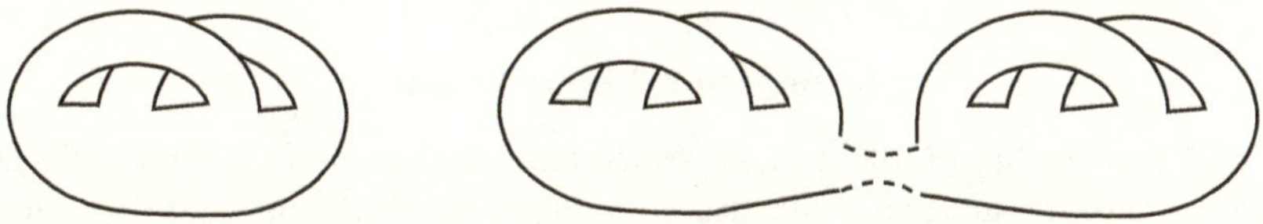

(a) (b)

Fig. 6.2. (a) genus 1, (b) genus *g*

affine structure. According to the classical immersion theory (see, e.g., [194]), an open parallelizable n-dimensional manifold can be immersed into $\mathbb{R}^n$. In our case, such an immersion is shown in Figure 6.2.

Such an immersion can be viewed as the developing map (6.2) with trivial monodromy.

Affine and projective structures naturally arise in the theory of completely integrable systems.

Example 6.1.8 Let (M^{2n}, ω) be a symplectic manifold with a Lagrangian foliation $\mathcal{L}^n$; cf. Appendix A.2. The leaves of $\mathcal{L}$ carry a canonical affine structure. To define this structure, we realize each leaf as a homogeneous $\mathbb{R}^n$-space.

Locally, the space of leaves of a Lagrangian foliation is a smooth manifold $N^n = M/\mathcal{L}$. Consider a point $x \in N$ and define an action of $\mathbb{R}^n = T_x^*N$ on the leaf L_x. Given a covector $p \in T_x^*N$, consider a smooth function f on N such that $\mathrm{d}f_x = p$. Lift f to M and consider its Hamiltonian vector field v. The field v is tangent to L_x and the restriction $v|_{L_x}$ does not depend on the choice of f. Any two such vector fields on L_x commute. Hence we have constructed a free action of $\mathbb{R}^n$ on L_x. This action is effective, and the leaf L_x is a homogeneous $\mathbb{R}^n$-space.

Consider the contact counterpart of this construction. Let M^{2n-1} be a contact manifold with a Legendrian foliation $\mathcal{L}^{n-1}$. Then the leaves of $\mathcal{L}$ carry a canonical projective structure. We define this structure in terms of a developing map to $\mathbb{RP}^{n-1}$.

Consider a leaf L_x. At each point of the leaf, one has the contact hyperplane, tangent to the leaf. Project this hyperplane to $T_x N$ where, as before, $N^n = M/\mathcal{L}$. A hyperplane in the tangent space is a contact element, that is, a point in $\mathbb{P}(T_x^* N) \cong \mathbb{RP}^n$. The constructed map $L_x \to \mathbb{RP}^n$ is an immersion, as follows from the complete non-integrability of the contact distribution.

Exercise 6.1.9 Prove the last assertion.

Monodromy as invariant

Let us discuss the problem of *classification* of projective structures. Under which condition are two projective structures on M diffeomorphic? In other words, what is the complete list of invariants of projective structures? This general problem is intractable in dimensions 3 and higher. The monodromy (6.3) is clearly a Diff(M)-invariant. The classification problem is then reduced to classification of projective structures with a fixed monodromy.

A *deformation*, or *homotopy*, is a continuous family of projective structures depending on one parameter. Let us consider deformations with fixed monodromy.

The following result can be considered an analog of Theorem 1.6.4.

Theorem 6.1.10 *Two projective structures on a compact manifold M which are homotopic with fixed monodromy are diffeomorphic.*

This theorem means that the monodromy is the unique continuous (or local) invariant of projective structures. We will prove Theorem 6.1.10 in Section 6.4.

Remark 6.1.11 Discrete invariants of projective structures are more difficult to classify. For instance, in the case $M = S^1$, one has a unique discrete invariant, namely the winding number; see Remark 1.6.6. The projective structures from Example 6.1.6 have the same monodromy but are not diffeomorphic if the corresponding 2-symbol words are different. Another important problem is to determine the homomorphisms (6.3) that correspond to projective structures.

Comment

The study of projective structures and, more generally, of G-structures on smooth manifolds was initiated by J. M. C. Whitehead [233] who proved the uniqueness of G-structures on simply connected manifolds, see also Ehresmann [59] for the case of a finite fundamental group. Ehresmann introduced the notion of developing map for G-structures. Example 6.1.6 is borrowed from [196]. Example 6.1.8, the Lagrangian case, is part of the Arnold–Liouville theorem in the theory of integrable systems; see [10, 15]. As to the Legendrian case, an adequate complete integrability theory is not yet available. The complete answer to the classification problem is known only in dimensions 1 and 2; see [39] and references therein for the two-dimensional case. See also [83] for a variety of results on projective and affine structures.

6.2 Projective structures in terms of differential forms

In this section we will reformulate the problem of existence of projective structures in terms of volume forms of a special type. In the odd-dimensional case, one relates projective structures to contact forms. The material of this and the next two sections is based on [164].

Volume forms

Let us introduce a kind of global coordinates for a projective structure. The definition is motivated by the notion of homogeneous coordinates on $\mathbb{RP}^n$.

Proposition 6.2.1 *A manifold M^n carries a projective structure if and only if there exist $n+1$ functions $f_1, \ldots, f_{n+1}$ on $\widetilde{M}$ satisfying the following two properties:*

(a) the space spanned by the functions $f_1, \ldots, f_{n+1}$ is invariant with respect to $\pi_1(M)$,

(b) the n-form

$$\Omega = \sum_{i=1}^{n+1} (-1)^{i+1} f_i \, \mathrm{d}f_1 \wedge \cdots \wedge \widehat{\mathrm{d}f_i} \wedge \cdots \wedge \mathrm{d}f_{n+1} \tag{6.6}$$

is a volume form on $\widetilde{M}$.

Proof Suppose there is a volume form given by (6.6) on $\widetilde{M}$. Denote by $\mathcal{F}$ the $(n+1)$-dimensional space spanned by the functions $f_1, \ldots, f_{n+1}$. To a point $x \in \widetilde{M}$ we assign the subspace $V_x \subset \mathcal{F}$ formed by the functions vanishing at x.

The dimension of the space V_x is n since the form Ω vanishes nowhere. Choosing an identification $\mathcal{F} \cong \mathbb{R}^{n+1}$, we obtain a map $\varphi : \widetilde{M} \to \mathbb{RP}^n$, defined up to a projective transformation (depending on the choice of the linear isomorphism $\mathcal{F} \cong \mathbb{R}^{n+1}$). Let us show that the map φ is a local diffeomorphism.

Since $\Omega \neq 0$, for any point x there is a function in $\mathcal{F}$ that is non-zero at x. Without loss of generality, we may assume that $f_{n+1}(x) \neq 0$. Let us put $x_i = f_i/f_{n+1}$ for $i = 1, \dots, n$.

Exercise 6.2.2 Check that, in a neighborhood of x, the form Ω is proportional to $\mathrm{d}x_1 \wedge \dots \wedge \mathrm{d}x_n$.

Thus the functions $x_1, \dots, x_n$ form a system of local coordinates on $\widetilde{M}$ in a neighborhood of x, so that φ is indeed a local diffeomorphism.

A different choice of the basis of the space $\mathcal{F}$ corresponds to a linear-fractional coordinate transformation so that one obtains a projective structure on $\widetilde{M}$. Furthermore, condition (a) insures that this projective structure descends to M.

Conversely, assume that M is equipped with a projective structure; let us construct the functions $f_1, \dots, f_{n+1}$. Let $\varphi : \widetilde{M} \to \mathbb{RP}^n$ be a developing map; it can be lifted to an immersion $\widetilde{\varphi} : \widetilde{M} \to \mathbb{R}^{n+1}$, defined up to homotheties. In homogeneous coordinates, one can write $\varphi(x) = (\varphi_1(x) : \dots : \varphi_n(x) : 1)$, and in linear coordinates on $\mathbb{R}^{n+1}$, one has

$$\widetilde{\varphi}(x) = (f(x)\,\varphi_1(x), \dots, f(x)\,\varphi_n(x), f(x))$$

where $f > 0$ is a function on $\widetilde{M}$. Choose an arbitrary volume form Ω on M, and thus on $\widetilde{M}$, and the standard volume form Ω_0 on $\mathbb{R}^{n+1}$. There is a unique function f such that $\Omega = i_{\widetilde{\varphi}(x)}\Omega_0$.

Exercise 6.2.3 Check that

$$f = \begin{vmatrix} \partial_1\varphi_1 & \dots & \partial_1\varphi_n \\ \dots & & \\ \partial_n\varphi_1 & \dots & \partial_n\varphi_n \end{vmatrix}^{-1/(n+1)} \tag{6.7}$$

where $\partial_i = \partial/\partial x^i$.

The pull-back of linear coordinates on $\mathbb{R}^{n+1}$ defines the functions $f_1, \dots, f_{n+1}$. This completes the proof. □

Remark 6.2.4 Proposition 6.2.1 provides a necessary and sufficient condition for the existence of a projective structure on M. However, the correspondence between projective structures and the space of functions $\langle f_1, \dots, f_{n+1}\rangle$ is not Diff(M)-invariant; to construct this space from a projective structure, we fixed a volume form on M.

Relation to contact geometry

Consider the case of an orientable odd-dimensional manifold, $\dim M = 2k - 1$. In this case, the existence of projective structures can be reformulated in terms of contact geometry; see Appendix A.2.

Corollary 6.2.5 *A manifold M^{2k-1} carries a projective structure if and only if there exist 2k functions $f_1, \ldots, f_k, g_1, \ldots, g_k$ on $\widetilde{M}$ satisfying the following two properties:*

(a) the space spanned by the functions $f_1, \ldots, f_k, g_1, \ldots, g_k$ is invariant with respect to $\pi_1(M)$,
(b) the 1-form

$$\alpha = \sum_{i=1}^{k} (f_i dg_i - g_i df_i) \tag{6.8}$$

is a contact form on $\widetilde{M}$.

Proof Given a contact form (6.8), one has $\alpha \wedge (\mathrm{d}\alpha)^{k-1} = \Omega$ as in Proposition 6.2.1, and hence M has a projective structure. Conversely, the existence of a projective structure on M implies the existence of a volume form (6.6). Split the set of functions involved into two subsets $f_1, \ldots, f_k, g_1, \ldots, g_k$ and consider the 1-form (6.8). Again, the relation $\alpha \wedge (\mathrm{d}\alpha)^{k-1} = \Omega$ implies that α is a contact form. □

Note that the choice of the contact structure is not unique: it depends on the choice of a Darboux basis.

6.3 Tensor densities and two invariant differential operators

In this section M^n is an arbitrary smooth manifold. We will define the notion of tensor densities on M which already played a prominent role in the one-dimensional case. We construct two invariant differential operators on the space of tensor densities. Although these operators are not directly related to projective structures we will use them in the next section to prove important results about projective structures.

Tensor densities on smooth manifolds

The space, $\mathcal{F}_\lambda(M)$, of tensor densities of degree $\lambda \in \mathbb{R}$ on M is the space of sections of the line bundle $(\wedge^n T^*M)^\lambda$. In local coordinates, a tensor density of degree λ can be expressed in the form

$$\phi = f(x_1, \ldots, x_n)(dx_1 \wedge \cdots \wedge dx_n)^\lambda.$$

The space $\mathcal{F}_\lambda(M)$ is a module over Diff(M) with the natural action. In local coordinates, it is given by the formula

$$T^\lambda_{g^{-1}}(\phi) = (J_g)^\lambda\, \phi(g), \qquad g \in \mathrm{Diff}(M)$$

where J_g is the Jacobian of g.

The Diff(M)-modules $\mathcal{F}_\lambda(M)$ and $\mathcal{F}_\mu(M)$ are isomorphic only if $\lambda = \mu$. Indeed, any isomorphism between these modules would be a unary invariant differential operator on tensor densities. Such operators are classified, and there is only one, the de Rham differential (cf. Comment in Section 2.1). The de Rham differential acts between tensor densities only in dimension 1 and then it has a kernel consisting of constants.

If M^n is closed and orientable then there is a Diff(M)-invariant pairing

$$\mathcal{F}_\lambda(M) \otimes \mathcal{F}_\mu(M) \to \mathbb{R} \tag{6.9}$$

with $\lambda + \mu = n$, already considered in Section 1.5 in the one-dimensional case.

Wronski operator

For every $\lambda_1, \ldots, \lambda_{n+1}$, there is an $(n+1)$-linear (skew-symmetric) invariant differential operator

$$W : \mathcal{F}_{\lambda_1}(M) \otimes \cdots \otimes \mathcal{F}_{\lambda_{n+1}}(M) \to \mathcal{F}_{\lambda+1}(M)$$

where $\lambda = \lambda_1 + \cdots + \lambda_{n+1}$. In local coordinates, this operator is given by the formula

$$W(\phi_1, \ldots, \phi_{n+1}) = \begin{vmatrix} \lambda_1 f_1 & \ldots & \lambda_{n+1} f_{n+1} \\ \partial_1 f_1 & \ldots & \partial_1 f_{n+1} \\ \vdots & \ddots & \vdots \\ \partial_n f_1 & \ldots & \partial_n f_{n+1} \end{vmatrix} (dx_1 \wedge \cdots \wedge dx_n)^{\lambda+1} \tag{6.10}$$

where $\phi_i = f_i\,(dx_1 \wedge \cdots \wedge dx_n)^{\lambda_i}$ for $i = 1, \ldots, n+1$ and $\partial_i f = \partial f/\partial x_i$.

Exercise 6.3.1 Check that formula (6.10) does not depend on the choice of local coordinates, so that the operator W is well-defined and commutes with the Diff(M)-action.

Note three particular cases.

Example 6.3.2 (a) In the one-dimensional case, the operator (6.10) is bilinear:

$$W(\phi_1, \phi_2) = (\lambda_1 f_1 f_2' - \lambda_2 f_1' f_2)(\mathrm{d}x)^{\lambda_1+\lambda_2+1}.$$

This is the well-known Schouten bracket. The Wronski determinant of two solutions of the Sturm–Liouville equation (see Section 1.6) is a very special particular case with $\lambda_1 = \lambda_2 = -1/2$.
(b) The volume form (6.6) provides a "regularized" version of the operator (6.10) in the case $\lambda_1 = \cdots = \lambda_{n+1} = 0$.
(c) If $\lambda_1 = \cdots = \lambda_{n+1} = -1/(n+1)$, then the operator W takes values in $C^\infty(M)$.

Invariant differential operators with values in Vect(M)

Let us define an n-linear (skew-symmetric) invariant differential operator

$$A : \mathcal{F}_{\lambda_1}(M) \otimes \cdots \otimes \mathcal{F}_{\lambda_n}(M) \to \mathrm{Vect}(M) \otimes_{C^\infty(M)} \mathcal{F}_{\lambda+1}(M)$$

where $\lambda = \lambda_1 + \cdots + \lambda_n$. In local coordinates this operator is given by the formula

$$A(\phi_1, \ldots, \phi_n) = \sum_{i=1}^{n} (-1)^i \begin{vmatrix} \lambda_1 f_1 & \cdots & \lambda_n f_n \\ \partial_1 f_1 & \cdots & \partial_1 f_n \\ \cdots & \cdots & \cdots \\ \widehat{\partial_i f_1} & \cdots & \widehat{\partial_i f_n} \\ \cdots & \cdots & \cdots \\ \partial_n f_1 & \cdots & \partial_n f_n \end{vmatrix} \frac{\partial}{\partial x_i} \otimes (\mathrm{d}x_1 \wedge \cdots \wedge \mathrm{d}x_n)^{\lambda+1} \tag{6.11}$$

Exercise 6.3.3 Check that the operator A is a well-defined Diff(M)-invariant differential operator, that is, formula (6.11) does not depend on the choice of local coordinates.

Remark 6.3.4 In the particular case $\lambda = -1$, the operator (6.11) takes values in Vect(M). In this case, it has a simple interpretation. Let ψ be a tensor density of degree μ. Then one has, for the Lie derivative,

$$L_{A(\phi_1,\ldots,\phi_n)}\psi = W(\phi_1, \ldots, \phi_n, \psi).$$

In this particular case, the above formula can be used for a definition of the operator (6.11).

We will consider one more invariant differential operator

$$\bar{A} : \mathcal{F}_{\lambda_1}(M) \otimes \cdots \otimes \mathcal{F}_{\lambda_{n+1}}(M) \to \mathrm{Vect}(M) \otimes_{C^\infty(M)} \mathcal{F}_{\lambda+1}(M)$$

with $\lambda = \lambda_1 + \cdots + \lambda_{n+1}$. This operator is defined as a composition of the operator A with multiplication, namely,

$$\bar{A}(\phi_1, \ldots, \phi_n; \phi_{n+1}) = \phi_{n+1}\, A(\phi_1, \ldots, \phi_n) \tag{6.12}$$

Example 6.3.5 In the one-dimensional case, the operator $\bar{A}$ is nothing else but the product of two tensor densities.

Relation between the operators $\bar{A}$ and W

Consider the special case

$$\lambda_1 + \cdots + \lambda_{n+1} = -1.$$

The operator $\bar{A}$ then takes values in $\mathrm{Vect}(M)$. In this case, there is a nice relation between the two invariant differential operators, $\bar{A}$ and W.

Exercise 6.3.6 For $\psi \in \mathcal{F}_\mu(M)$, check the identity for the Lie derivative:

$$L_{\bar{A}(\phi_1,\ldots,\phi_n;\phi_{n+1})}\psi = \phi_{n+1}\, W(\phi_1, \ldots, \phi_n, \psi) + \mu\, \psi\, W(\phi_1, \ldots, \phi_{n+1}). \tag{6.13}$$

6.4 Projective structures and tensor densities

In this section we will give one more equivalent definition of projective structure. Our approach is similar to the classical theory in the one-dimensional case; see Section 1.3. We associate with each projective structure $\mathfrak{P}$ on M an $(n+1)$-dimensional space $\mathcal{F}_\mathfrak{P}$ of tensor densities of degree $-1/(n+1)$ on $\widetilde{M}$ such that the operator W defines a volume form on this space. In the one-dimensional case, $\mathcal{F}_\mathfrak{P}$ is just the space of solutions of a Sturm–Liouville equation.

Projective structures and tensor densities

The operator (6.10) is closely related to the notion of projective structure. Our idea (similar to the one-dimensional case) consists in assigning a tensor weight to the functions $f_1, \ldots, f_{n+1}$ from Proposition 6.2.1.

The following statement is a refined, $\mathrm{Diff}(M)$-invariant version of Proposition 6.2.1.

Theorem 6.4.1 *There is a one-to-one* $\mathrm{Diff}(M)$*-invariant correspondence* $\mathfrak{P} \mapsto \mathcal{F}_{\mathfrak{P}}$ *between projective structures on* M *and* $(n+1)$*-dimensional subspaces of* $\mathcal{F}_{-1/(n+1)}(\widetilde{M})$ *satisfying the following two properties:*

(a) the space $\mathcal{F}_{\mathfrak{P}}$ *is invariant with respect to* $\pi_1(M)$,
(b) for every basis $\phi_1, \ldots, \phi_{n+1}$ *in* $\mathcal{F}_{\mathfrak{P}}$, *one has*

$$W(\phi_1, \ldots, \phi_{n+1}) = const \neq 0. \tag{6.14}$$

Proof Given a space $\mathcal{F}_{\mathfrak{P}} \subset \mathcal{F}_{-1/(n+1)}(\widetilde{M})$ satisfying conditions (a) and (b), for every point x of $\widetilde{M}$ we consider the subspace $\mathcal{V}_x \subset \mathcal{F}_{\mathfrak{P}}$ consisting of tensor densities, vanishing at x. This defines a developing map of a projective structure. The proof of this fact is similar to that of the first part of Proposition 6.2.1.

Conversely, given a projective structure $\mathfrak{P}$ on M, fix an arbitrary volume form Ω. One then obtains an $(n+1)$-dimensional space of functions $\langle f_1, \ldots, f_{n+1}\rangle$ on $\widetilde{M}$ (see Proposition 6.2.1). For every function f from this space, the tensor density $f\,\Omega^{-1/(n+1)}$ is $\mathrm{Diff}(M)$-invariant. This immediately follows from formula (6.7).

Exercise 6.4.2 Check that the construction is independent of the choice of the volume form Ω.

Hence the result. □

Reconstructing the $\mathrm{sl}(n+1, \mathbb{R})$-action

Let us recall that, in the one-dimensional case, the action of $\mathrm{sl}(2, \mathbb{R})$, corresponding to a projective structure, was reconstructed from the respective Sturm–Liouville equation via products of pairs of its solutions; see Exercise 1.3.6. Similarly, in the multi-dimensional case, we wish to recover the $\mathrm{sl}(n+1, \mathbb{R})$-action, corresponding to a projective structure $\mathfrak{P}$, from the $(n+1)$-dimensional space $\mathcal{F}_{\mathfrak{P}}$ of tensor densities constructed in Theorem 6.4.1.

Proposition 6.4.3 *Let* $\mathfrak{P}$ *be a projective structure on a manifold* M. *To reconstruct the* $\mathrm{sl}(n+1, \mathbb{R})$*-action on* $\widetilde{M}$ *it suffices to take the image of the operator*

$$\bar{A} : \Lambda^n \mathcal{F}_{\mathfrak{P}} \otimes \mathcal{F}_{\mathfrak{P}} \to \mathrm{Vect}(\widetilde{M}).$$

This image is a Lie subalgebra of $\mathrm{Vect}(\widetilde{M})$, *isomorphic to* $\mathrm{sl}(n+1, \mathbb{R})$.

Proof Let $\mathfrak{a}$ be the image of $\bar{A}|_{\Lambda^n \mathcal{F}_{\mathfrak{P}} \otimes \mathcal{F}_{\mathfrak{P}}}$. Identity (6.13) implies that $\mathfrak{a}$ is a Lie subalgebra of $\mathrm{Vect}(\widetilde{M})$. Indeed, the commutator of two vector fields of the form

$\bar{A}(\phi_1, \ldots, \phi_n; \phi_{n+1})$, where $\phi_i \in \mathcal{F}_{\mathfrak{P}}$, can be computed by the Leibnitz rule, using the fact that the operator W is constant on the elements of $\mathcal{F}_{\mathfrak{P}}$.

Identity (6.13) also implies that the Lie algebra $\mathfrak{a}$ acts on the space $\mathcal{F}_{\mathfrak{P}}$. Furthermore, this action preserves the volume form on $\mathcal{F}_{\mathfrak{P}}$, defined by the operator W, since this operator is invariant. One therefore has a homomorphism

$$\mathfrak{a} \to \mathrm{sl}(n+1, \mathbb{R}).$$

For a point $x \in M$, we may assume, without loss of generality, that $\phi_{n+1} \neq 0$. We then fix a coordinate system in a neighborhood of this point: $x_i = \phi_i / \phi_{n+1}$.

Exercise 6.4.4 Check that, in the chosen coordinate system, the Lie algebra $\mathfrak{a}$ is generated by the vector fields (6.5).

Hint. It is almost a tautology that, in the chosen coordinate system, the elements of $\mathcal{F}_{\mathfrak{P}}$ are written as $\phi = f\,(\mathrm{d}x_1 \wedge \cdots \wedge \mathrm{d}x_n)^{-1/(n+1)}$ where f is a polynomial of degree ≤ 1 in x.

Therefore the image of this homomorphism coincides with $\mathrm{sl}(n+1, \mathbb{R})$.

Finally, let us show that the above homomorphism of Lie algebras has no kernel. The operator $\bar{A}$ is defined on the space $\Lambda^n \mathcal{F}_{\mathfrak{P}} \otimes \mathcal{F}_{\mathfrak{P}}$ of dimension $(n+1)^2$. The following exercise shows that the operator $\bar{A}$ has a non-trivial kernel.

Exercise 6.4.5 Check that if $\phi_1, \ldots, \phi_{n+1} \in \mathcal{F}_{\mathfrak{P}}$ then

$$\mathrm{Alt}_{1,\ldots,n+1}\, \bar{A}(\phi_1, \ldots, \phi_n; \phi_{n+1}) = 0$$

where Alt means complete anti-symmetrization.

It follows that $\dim \mathfrak{a} \leq n^2 + 2n$.

We proved that the vector fields of the form $\bar{A}(\phi_1, \ldots, \phi_n; \phi_{n+1})$ with $\phi_i \in \mathcal{F}_{\mathfrak{P}}$ span a subalgebra of $\mathrm{Vect}(\widetilde{M})$ isomorphic to $\mathrm{sl}(n+1, \mathbb{R})$ and acting on the subspace $\mathcal{F}_{\mathfrak{P}}$. Hence this Lie algebra preserves the projective structure $\mathfrak{P}$. Proposition 6.4.3 is proved. □

Proof of Theorem 6.1.10

Our proof goes along the same lines as that of Theorem 1.6.4.

Let us use the homotopy method. Consider a family $\mathfrak{P}_t$ of projective structures on a compact manifold M. We think of a projective structure $\mathfrak{P}$ as a space of $-1/(n+1)$-densities $\mathcal{F}_{\mathfrak{P}}$. Fix a basis $\phi_{1_t}, \ldots, \phi_{n+1_t}$ in $\mathcal{F}_{\mathfrak{P}_t}$ so that

$W(\phi_{1_t}, \ldots, \phi_{n+1_t}) \equiv 1$ and the monodromy representation expressed in this basis is also independent of t. We will prove that for every t there exists a diffeomorphism g_t that takes the basis $\phi_{1_0}, \ldots, \phi_{n+1_0}$ to $\phi_{1_t}, \ldots, \phi_{n+1_t}$.

According to the homotopy method, it suffices to find a family of vector fields X_t on M such that their lift $\widetilde{X}_t$ to $\widetilde{M}$ satisfies

$$\dot{\phi}_{i_t} = L_{\widetilde{X}_t}(\phi_{i_t}) \tag{6.15}$$

where dot is the derivative with respect to t and $i = 1, \ldots, n+1$. Let us simplify the notation and suppress the subscript t everywhere.

Lemma 6.4.6 *The solution of the homotopy equation (6.15) is the vector field*

$$\widetilde{X} = \mathrm{Alt}_{1,\ldots,n+1}\, \bar{A}(\phi_1, \ldots, \phi_n; \dot{\phi}_{n+1}). \tag{6.16}$$

Proof In local coordinates, $\widetilde{X} = \sum_j h^j \partial/\partial x_j$, and the equation (6.15) is a system of $(n+1)$ linear equations

$$\dot{\phi}_i = \sum_{j=1}^{n} \left(h^j \frac{\partial \phi_i}{\partial x_j} - \frac{1}{n+1} \frac{\partial h^j}{\partial x_j} \phi_i \right)$$

in the variables h^j with $j = 1, \ldots, n$ and $\zeta = \sum_j \partial h^j/\partial x_j$ which we temporarily consider as an independent variable.

Exercise 6.4.7 (a) Using the fact that $W(\phi_1, \ldots, \phi_{n+1}) \equiv 1$, check that the solution of the above system of linear equations is

$$h^i = (-1)^i \begin{vmatrix} \dot{f}_1 & \ldots & \dot{f}_{n+1} \\ f_1 & \ldots & f_{n+1} \\ \partial_1 f_1 & \ldots & \partial_1 f_{n+1} \\ \ldots & & \\ \widehat{\partial_i f_1} & \ldots & \widehat{\partial_i f_{n+1}} \\ \ldots & & \\ \partial_n f_1 & \ldots & \partial_n f_{n+1} \end{vmatrix}, \qquad \zeta = -(n+1) \begin{vmatrix} \dot{f}_1 & \ldots & \dot{f}_{n+1} \\ \partial_1 f_1 & \ldots & \partial_1 f_{n+1} \\ \ldots & & \\ \partial_i f_1 & \ldots & \partial_i f_{n+1} \\ \ldots & & \\ \partial_n f_1 & \ldots & \partial_n f_{n+1} \end{vmatrix}$$

(b) Using the fact that $\dot{W}(\phi_1, \ldots, \phi_{n+1}) \equiv 0$, check that ζ, given by the last formula, indeed equals $\sum_j \partial h^j/\partial x_j$.

Expanding the determinant in the above formula for h^i in the first row, we obtain formula (6.16). This completes the proof of the lemma. □

To complete the proof of the theorem note that the vector field $\widetilde{X}$ descends to M. Indeed, let $\gamma \in \pi_1(M)$. It follows from the determinantal formula for h^i from Exercise 6.4.7 that $\gamma^* h^i = h^i \cdot \det T_\gamma$ where γ is understood as a diffeomorphism of $\widetilde{M}$. Finally, $T_\gamma \in \mathrm{SL}(n+1, \mathbb{R})$ and therefore $\gamma^* h^i = h^i$.

Theorem 6.1.10 is proved.

Remark 6.4.8 If the "infinitesimal variation" $\dot{\phi}_1, \ldots, \dot{\phi}_{n+1}$ belongs to the space $\mathcal{F}_{\mathfrak{P}}$, then the vector field (6.16) vanishes (cf. Exercise 6.4.5).

Multi-dimensional Sturm theorem on zeroes

Let M^n be a simply connected manifold with a projective structure $\mathfrak{P}$ and φ a developing map. A *geodesic* on M is the preimage of a projective line $\mathbb{RP}^1 \subset \mathbb{RP}^n$ under φ. Note that geodesics can be disconnected. A *geodesic submanifold* is characterized by the following property: if a geodesic in M is tangent to the submanifold then this geodesic lies entirely in this submanifold. Equivalently, a geodesic submanifold of dimension k is the preimage of $\mathbb{RP}^k \subset \mathbb{RP}^n$.

The space $\mathcal{F}_{\mathfrak{P}}$ is useful for description of geodesic submanifolds of M. Let $V \subset \mathcal{F}_{\mathfrak{P}}$ be a k-dimensional subspace. The condition $\phi = 0$ for all $\phi \in V$ determines a geodesic submanifold of M of codimension k and, conversely, every geodesic submanifold corresponds to a subspace in V.

We have already mentioned the classic Sturm theorem on zeroes at the end of Section 1.3. Let us discuss its multi-dimensional analog. We consider here only the case $\dim M = 2$ leaving higher dimensions to the reader's imagination.

Theorem 6.4.9 *Let M^2 be a simply connected surface with a projective structure and $\gamma_1, \gamma_2, \gamma_3$ three geodesics. Then the intersection points of each connected component of γ_1 with γ_2 and γ_3 alternate (see Figure 6.3).*

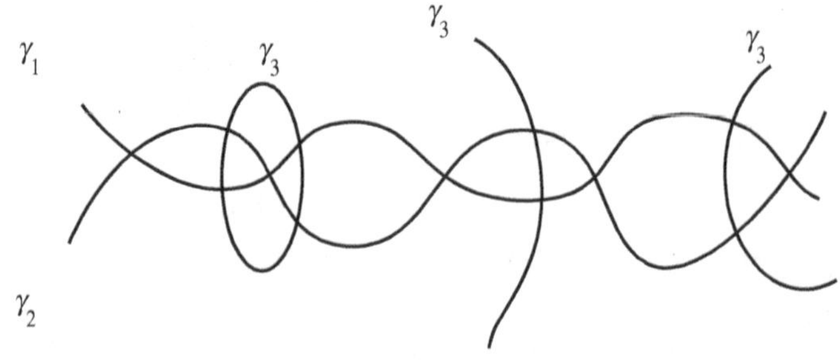

Fig. 6.3. Multi-dimensional Sturm theorem

Proof The geodesics γ_1, γ_2 and γ_3 are zeroes of some densities ϕ_1, ϕ_2 and ϕ_3. Choosing an area form Ω on M, let us consider the functions f_i, $i = 1, 2, 3$, such that $\phi_i = f_i\, \Omega^{-1/3}$. Consider the 1-form $f_2\, \mathrm{d} f_3 - f_3\, \mathrm{d} f_2$. We claim that its restriction on γ_1 does not vanish.

Indeed, let t be a parameter on a connected component of γ_1. Consider a small neighborhood U of γ_1. One can use t and f_1 as coordinates in U. The Wronski determinant

$$W = \begin{vmatrix} f_1 & f_2 & f_3 \\ 0 & \frac{\partial f_2}{\partial t} & \frac{\partial f_3}{\partial t} \\ 1 & \frac{\partial f_2}{\partial f_1} & \frac{\partial f_3}{\partial f_1} \end{vmatrix} \neq 0.$$

On γ_1 one has $f_1 = 0$ and hence

$$W = \begin{vmatrix} f_2 & f_3 \\ \frac{\partial f_2}{\partial t} & \frac{\partial f_3}{\partial t} \end{vmatrix} \neq 0.$$

It follows that $f_2\, \mathrm{d} f_3 - f_3\, \mathrm{d} f_2 = W \mathrm{d}t$ does not vanish and defines a projective structure on γ_1. Therefore zeroes of the functions $f_2(t)$ and $f_3(t)$ alternate by the classic Sturm theorem. $\square$

6.5 Moduli space of projective structures in dimension 2, by V. Fock and A. Goncharov

The space of projective structures on Riemann surfaces was studied mainly by S.Choi and W.Goldman [39, 40], F.Labourie [127], J.Loftin [139] and, implicitly, N.Hitchin [97]. Given a projective structure on a Riemann surface Σ, consider the developing map $\phi : \tilde{\Sigma} \to \mathbb{RP}^2$ where $\tilde{\Sigma}$ is a universal cover of Σ; the map ϕ is defined up to the right action of PGL(3, $\mathbb{R}$). A projective structure on Σ is called *convex* if ϕ is an embedding and its image is a convex domain.

If the surface Σ has boundary, the space of projective structures on Σ is obviously infinite dimensional even with convexity requirement. Indeed, any projective invariant of the image of any subsegment of the boundary under ϕ is an invariant of the projective structure. In order to make the space of projective structures more similar to the one given by Theorem 6.1.10, let us impose a certain requirement on the behavior of the projective structure at the vicinity of the boundary. Namely we require that the projective structure is extendable to the boundary, and the image of the boundary under the developing map ϕ belongs to a line. It follows from convexity that, in the non-degenerate case, the image of a boundary component forms a segment, connecting fixed points of the monodromy operator. The boundary is then called *linear*. Otherwise, it is the limiting case of the previous one, when the fixed points coincide. The latter case is called degenerate.

A *framing* of the projective structure is an orientation of all non-degenerate boundary components. The space of framed surfaces is a covering of the space of non-framed ones, ramified over surfaces with degenerate projective structures. The degree of the covering is less than or equal to 6^s, where s is the number of holes, since for every hole there are two framings and, in general, three choices of pairs of eigenvectors of the monodromy operator.

The object of main interest for us in this section is the space of framed convex real projective structures on an oriented surface Σ with linear boundary and with oriented boundary components, considered up to the group $\mathrm{Diff}_0(\Sigma)$ of diffeomorphisms, homotopy equivalent to the identity. We denote this space by $\mathcal{T}_3^H(\Sigma)$ (the index 3 indicates that we are dealing with the group $\mathrm{PGL}(3, \mathbb{R})$ and H stands for holes.)

There exist the following canonical maps:

- The map I of the ordinary Teichmüller space $\mathcal{T}_2^H(\Sigma)$ to $\mathcal{T}_3^H(\Sigma)$. Recall that $\mathcal{T}_2^H(\Sigma)$ is the space of complex structures on Σ, considered up to the action of $\mathrm{Diff}_0(\Sigma)$. Indeed, as a result of the Poincaré uniformization theorem, any complex surface can be represented as a quotient of the upper half plane by a discrete group. Consider the upper half plane as a model of the hyperbolic plane and replace it by the projective model (interior of a conic in $\mathbb{RP}^2$): one obtains a canonical projective structure on the quotient; cf. Example 6.1.4. The orientations of holes are induced trivially.
- The map μ from $\mathcal{T}_3^H(\Sigma)$ to the space $\mathcal{M}_3(\Sigma)$ of homomorphisms of $\pi_1(\Sigma)$ to $\mathrm{PGL}(3, \mathbb{R})$ with discrete image, considered up to conjugation. The results of F. Labourie [127], S. Choi and W. Goldman [40], based on the work of N. Hitchin [97], show that the image of this map is a connected component of $\mathcal{M}_3(\Sigma)$ and, moreover, given a complex structure on Σ, this space is isomorphic to the space of holomorphic cubic differentials on Σ. The space $\mathcal{M}_3(\Sigma)$ possesses a canonical Atiyah–Hitchin Poisson structure. Since the map μ is a local diffeomorphism, it induces a Poisson structure on $\mathcal{T}_3^H(\Sigma)$.
- The involution $\sigma : \mathcal{T}_3^H(\Sigma) \to \mathcal{T}_3^H(\Sigma)$, defined by the property that $\phi(\sigma x)$ is projectively dual to $\phi(x)$ for $x \in \mathcal{T}_3^H(\Sigma)$. The map σ is a Poisson map.
- The action of the *mapping class group*:

$$\mathcal{T}_3^H(\Sigma) \times (\mathrm{Diff}(\Sigma)/\mathrm{Diff}_0(\Sigma)) \to \mathcal{T}_3^H(\Sigma),$$

  where $\mathrm{Diff}_0(\Sigma)$ is the connected component of $\mathrm{Diff}(\Sigma)$. This action preserves the Poisson structure.

Now we will give a set of global parameterizations of $\mathcal{T}_3^H(\Sigma)$, describe its natural Poisson structure in terms of these parameterizations and give explicit formulæ for the action of the mapping class group.

A model problem

Let us first study a simpler problem which, in a way, contains most of the tools used.

Definition 6.5.1 Let $\mathcal{P}_3^n$ be the space of pairs of convex n-gons in $\mathbb{RP}^2$, inscribed one into another, considered up to projective transformations.

This space is a discrete approximation to the space of parameterized closed strictly convex curves in $\mathbb{RP}^2$. Fix a set R of n points on the standard circle S^1, then with a convex curve $\gamma : S^1 \to \mathbb{RP}^2$ one associates the convex polygon with vertices $\gamma(R)$ and the polygon with edges tangent to γ at $\gamma(R)$.

The space $\mathcal{P}_3^n$ is a Poisson manifold and there are analogs of the maps μ, σ, I and the mapping class group action.

The natural map $\mu : \mathcal{P}_3^n \to F_3^n/\mathrm{PGL}(3, \mathbb{R})$ is an analog of the map μ for $\mathcal{T}_3^H(\Sigma)$. The projective duality interchanges inscribed and circumscribed polygons and acts as an involution σ of $\mathcal{P}_3^n$. There exists a canonical map $I : \mathcal{P}_2^n \to \mathcal{P}_3^n$, where $\mathcal{P}_2^n$ is the space of polygons inscribed into a conic. Indeed, to such a polygon one canonically assigns the circumscribed polygon formed by the tangents to the conic at the vertices of the original polygon.

Exercise 6.5.2 Show that the image of I coincides with the set of fixed points of σ.

Hint. Compare with Exercise 1.4.5 (b).

The role of the mapping class group is played by the cyclic group, permuting the vertices of the polygons.

Let us now proceed to the parameterization of the spaces $\mathcal{P}_3^n$. Cut the inscribed polygons into triangles by diagonals and mark two distinct points on every edge of the triangulation, except the edges of the polygon, and mark also one point inside each triangle; see Figures 6.4 and 6.5 below.

Theorem 6.5.3 *There exists a canonical bijective correspondence between the space $\mathcal{P}_3^n$ and assignments of positive real numbers to the marked points.*

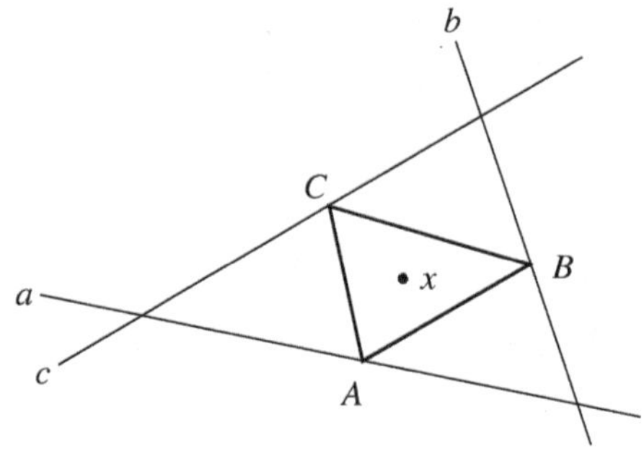

Fig. 6.4. The moduli space $\mathcal{P}_3^3$

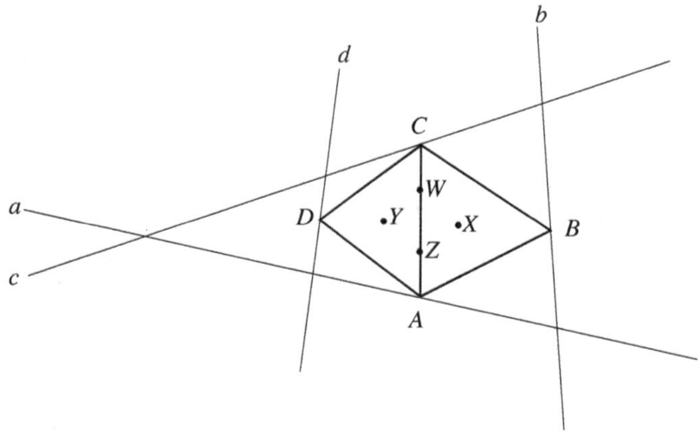

Fig. 6.5. The moduli space $\mathcal{P}_3^4$

Proof The proof is constructive: we will describe how to construct numbers from a pair of polygons and vice versa.

We start with a remark about notation. We denote points, resp. lines, by capital, resp. lower case, letters. A triangle in $\mathbb{RP}^2$ is determined by neither its vertices nor by its sides, since there exist four triangles for each triple of vertices or sides. If a triangle is shown on a figure, it is clear which one corresponds to the vertices, since only one of four fits entirely into the figure. If we want to indicate a triangle which does not fit, we add to its vertices, in parentheses, a point which belongs to the interior triangle. For example, points A, B and C on Figure 6.4 are vertices of the triangles ABC, $ABC(a \cap c)$, $ABC(a \cap b)$ and $ABC(b \cap c)$.

We will also use a definition of the cross-ratio, different from the one in Section 1.2. Namely, by the cross-ratio of points A, B, C, D we mean the value at D of a projective coordinate which is equal to ∞ at A, -1 at B and 0 at C; cf. Exercise 1.2.1 (b).

Exercise 6.5.4 Check that the defined cross-ratio is as follows:

$$[A, B, C, D] = \frac{(A - B)(C - D)}{(A - D)(B - C)}.$$

Let us start with the case of $\mathcal{P}_3^3$. This is the space of pairs of triangles abc and ABC (Figure 6.4), where the latter is inscribed into the former, and considered up to projective transformations. Since projective transformations act transitively on quadruples of non-collinear points, $\mathcal{P}_3^3$ is one-dimensional. A unique invariant of a pair is called the *triple ratio*, namely, the ratio

$$X = \frac{|A\,(a \cap c)||B\,(b \cap a)||C\,(c \cap b)|}{|A\,(a \cap b)||B\,(b \cap c)||C\,(c \cap a)|}$$

where the distances are measured in any Euclidean metric in $\mathbb{R}^2$.

Exercise 6.5.5 Show that the triple ratio is a projective invariant.

Hint. If the lines $A(b \cap c)$, $B(c \cap a)$ and $C(a \cap b)$ are concurrent then the triple ratio equals 1: this is the Ceva theorem of elementary geometry. In general, intersect the lines $B(c \cap a)$ and $C(a \cap b)$, connect their intersection to the point $(b \cap c)$, intersect this line with the line $(c \cap a)(a \cap b)$, and consider the cross-ratio of the four points on the latter line.

The fact that ABC is indeed inscribed into abc implies that X is positive.

Now consider the next case, $\mathcal{P}_3^4$. Likewise, this is the space of pairs of quadrilaterals $abcd$ and $ABCD$, where the latter is inscribed into the former, and considered up to projective transformations. The space of such configurations has dimension 4. Two parameters of these configurations are given by the triple ratio X of the triangle ABC, inscribed into abc, and the triple ratio Y of ACD, inscribed into acd.

Two other parameters are given by the cross-ratios of quadruples of points $(a \cap c)(a \cap d)A(a \cap b)$ and $(c \cap a)(c \cap d)C(c \cap b)$, denoted by Z and W, respectively. They are defined since the points of each of these quadruples are collinear. Note that Z and W are positive (cf. Exercise 6.5.4). Assign the coordinates X, Y, Z and W to the marked points, as shown in Figure 6.5.

Finally let us consider the space $\mathcal{P}_3^n$ for an arbitrary n. A point of this space is represented by a polygon $A_1 \ldots A_n$, inscribed into $a_1 \ldots a_n$. Cut the polygon $A_1 \ldots A_n$ into triangles by diagonals. Given a triangle $A_i A_j A_k$ of the triangulation, inscribed into $a_i a_j a_k$, assign to the point inside it the respective triple ratio. Likewise, given a pair of adjacent triangles $A_i A_j A_k$ and $A_j A_k A_l$, forming a quadrilateral $A_i A_j A_k A_l$ inscribed into the quadrilateral $a_i a_j a_k a_l$, one finds the respective pair of cross-ratios and assigns it to two points on the diagonal A_jA_k, the cross-ratio of the points on the line a_j being assigned to the point closer to the point A_j. This completes the construction. □

Parameterization of projective structures

Now let us proceed to the analogous statements about projective structures on surfaces. Let Σ be a Riemann surface of genus g with s boundary components. Assume that $s \geq 1$ and, moreover, if $g = 0$ then $s \geq 3$. Shrink all the boundary components to points. Then the surface Σ can be cut into triangles with vertices at the contracted boundary components. Let us assign two distinct marked points to each edge of the triangulation and one marked point to the centre of every triangle.

Theorem 6.5.6 *There exists a canonical bijective correspondence between the space $\mathcal{T}_3^H(\Sigma)$ of framed convex real projective structures on Σ and assignments of positive real numbers to the marked points.*

Proof The argument is analogous to the proof of Theorem 6.5.3.

Let us first construct a surface, starting from a collection of positive real numbers, assigned to a triangulation. Consider the universal cover $\tilde{\Sigma}$ of the surface and lift the triangulation, along with the marked points and numbers, from Σ to $\tilde{\Sigma}$. As described above, with any finite subpolygon of the arising infinite triangulated polygon we can associate a pair of polygons in $\mathbb{RP}^2$, one inscribed into another. Consider the union U of all inscribed polygons, corresponding to such finite subpolygons. The group $\pi_1(\Sigma)$ acts naturally on $\tilde{\Sigma}$ and thus acts on U by projective transformations. The desired projective surface is $U/\pi_1(\Sigma)$.

Note that one can also consider the intersection of all circumscribed polygons. Its quotient by the fundamental group, in general, gives another projective structure on Σ.

Now let us describe the construction of numbers from a given framed convex projective structure and a triangulation. Take a triangle and send it to $\mathbb{RP}^2$ by a developing map. The vertices of the triangle correspond to boundary components. Let the map ϕ be chosen. Given a boundary component S, the framing allows us to define a canonical flag (A, a) on $\mathbb{RP}^2$, invariant under the action of the monodromy operator around S. If the boundary component is degenerate, such a flag is uniquely characterized by the requirement that A belongs to the limit of the image under ϕ of a point, tending to S. If the boundary component is non-degenerate, we take the line containing the image of S for a, and one of the endpoints on the image of ϕ for A. The choice between the two endpoints is given by the framing.

Associating flags to all three vertices of the triangle, one obtains a point of $\mathcal{P}_3^3$. The corresponding coordinate is assigned to the central marked point of the original triangle. Similarly, considering two adjacent triangles of the triangulation, one constructs the numbers for the marked points on the edges. □

Let us summarize the properties of the constructed coordinates both for $\mathcal{P}_3^3$ and $\mathcal{T}_3^H(\Sigma)$:

(1) Poisson bracket. It turns out, that the Poisson brackets between the coordinates are very simple to describe, namely,

$$\{X^i, X^j\} = \varepsilon^{ij} X^i X^j$$

where ε^{ij} is a skew-symmetric matrix with integer entries. To define the matrix ε^{ij}, consider the graph with vertices in marked points and oriented edges connecting them, as shown in Figure 6.6 (we show edges connecting marked points in one triangle only; points of other triangles are connected by arrows in the same way). Then

$$\varepsilon^{ij} = (\text{number of arrows from } i \text{ to } j) - (\text{number of arrows from } j \text{ to } j).$$

To prove this formula one needs to compare it with any other expression for the Atyah–Hitchin Poisson structure. The most appropriate, from our point of view, is the expression using classical r-matrices in [70], but still the verification is technically complicated. However, it is an easy exercise to verify that this bracket does not depend on the triangulation, and therefore it can be taken for a definition.

(2) Once we have positive numbers assigned to marked points, the construction of the corresponding projective surface is explicit. In particular, one can compute the monodromy group of the corresponding projective structure or, in other words, the image of the map μ.

We describe the answer explicitly in the following graphical way. Starting from the triangulation, construct a graph, embedded into the

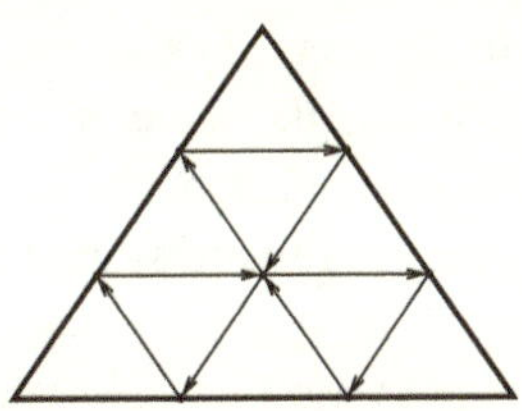

Fig. 6.6. Poisson structure tensor

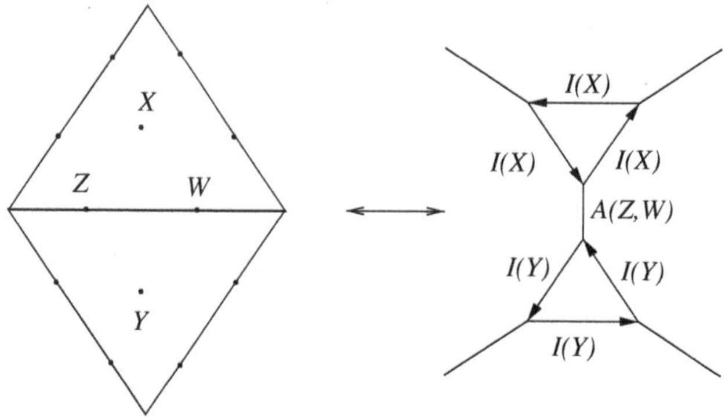

Fig. 6.7. Construction of the monodromy group

surface, by drawing edges transversal to the sides of the triangles (the dual graph), and inside each triangle connect the ends of edges pairwise by three more edges, as shown in Figure 6.7. Orient these edges counterclockwise. Now assign to each edge an element of PGL(3, $\mathbb{R}$), constructed from the numbers as shown in Figure 6.7. Here

$$I(X) = \begin{pmatrix} 0 & 0 & 1 \\ 0 & -1 & -1 \\ X & X+1 & 1 \end{pmatrix} \text{ and } A(Z, W) = \begin{pmatrix} 0 & 0 & Z^{-1} \\ 0 & -1 & 0 \\ W & 0 & 0 \end{pmatrix}.$$

With a closed path in the graph one associates an element of PGL(3, $\mathbb{R}$) by taking the product of these group elements (or their inverses, if the path goes against orientation), assigned to the consecutive edges. The image of the fundamental group of the graph is the desired monodromy group.

The proof of this statement is also constructive. Once we have a configuration of flags from $\mathcal{P}_3^3$ with triple ratio X, one can fix a coordinate system in $\mathbb{RP}^2$. Namely, take a coordinate system where the points $b \cap c$, A, B and C have the following coordinates

$$[0:1:0], \qquad [1:-1:1], \qquad [0:0:1] \quad \text{and} \quad [1:0:0],$$

respectively (see Figure 6.4). The line a has coordinates $[1 : 1 + X : X]$. The cyclic permutation of the flags changes the coordinate system by the matrix $I(X)$. Likewise, given a quadruple of flags (F_1, F_2, F_3, F_4) with two cross-ratios Z and W, the coordinate system associated with the triple (F_2, F_4, F_1) is related to the coordinate system associated with the triple (F_4, F_2, F_3) by the matrix $A(Z, W)$.

(3) The involution σ acts in a very simple way shown in Figure 6.8, where $Z' = \frac{W(1+Y)}{Y(1+X)}$ and $W' = \frac{Z(1+X)}{X(1+Y)}$. In particular, a point of $\mathcal{T}_3^H(\Sigma)$ is fixed by

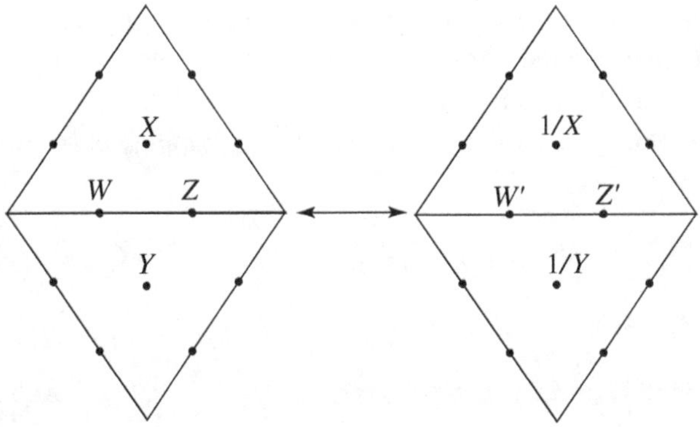

Fig. 6.8. Involution σ

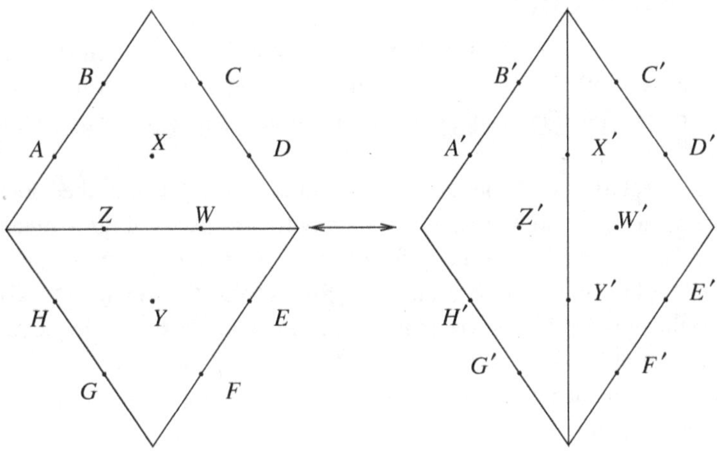

Fig. 6.9. Flip

σ if the two coordinates on each edge coincide and the coordinates in the centre of each triangle are equal to 1. Taking into account that the set of σ-stable points is just the ordinary Teichmüller space, one obtains its parameterization. This parameterization coincides with the one described in [68].

(4) Each triangulation of Σ provides its own coordinate system and, in general, the transition from one such system to another is given by complicated rational maps. However, any change of triangulation may be decomposed into a sequence of elementary changes, the so-called *flips*. A flip removes an edge of the triangulation and inserts another one into the arising quadrilateral, as shown on Figure 6.9. This figure also shows how

> the numbers at the marked points change under the flip. Note that these formulæ, in particular, make it possible to pass from a triangulation to itself, but transformed by a non-trivial element of the mapping class group of Σ, and thus give explicit formulae for the mapping class group action.

Here one has

$$A' = A\,\tfrac{Z}{1+Z}, \qquad B' = B\,\tfrac{XW(1+Z)}{1+W+WX+WXZ}, \qquad C' = C\,\tfrac{1+W+WX+WXZ}{1+W},$$

$$D' = D(1+W), \qquad E' = E\,\tfrac{W}{1+W}, \qquad F' = F\,\tfrac{YZ(1+W)}{1+Z+ZY+ZYW},$$

$$G' = G\,\tfrac{1+Z+ZY+ZYW}{1+Z}, \qquad H' = H(1+Z), \qquad X' = \tfrac{1+W}{XW(1+Z)},$$

$$Y' = \tfrac{1+Z}{YZ(1+W)}, \qquad Z' = \tfrac{1+W+WX+WXZ}{XZW(1+Z+ZY+ZWY)},$$

$$W' = \tfrac{1+Z+ZY+ZWY}{YZW(1+W+WX+WXZ)}.$$

These formulæ can be derived directly or, more simply, using the cluster algebras technique; see [69]. Note that the resulting rational functions have positive integer coefficients. One can show that this remains true for any composition of flips.

Let us conclude with remarks about generalizations. The described construction can be generalized almost automatically to the spaces of projective structures on surfaces whose boundary is not linear but polygonal. More precisely, every point of the boundary is projectively isomorphic to a neighborhood of the origin in the upper half-plane or in the positive octant. On each linear boundary segment a point is chosen. Obviously, $\mathcal{T}_3^H(\Sigma)$ and $\mathcal{P}_3^n$ are both particular cases of this more general setting. We leave this generalization as an exercise for the industrious reader.

Another generalization may be obtained by replacing $\mathrm{PGL}(3,\mathbb{R})$ by any other split simple Lie group [69]. Although one can define the corresponding analogs of the space $\mathcal{T}_3^H(\Sigma)$, it is unknown whether it can be interpreted as the moduli space of certain local geometric structures.

7
Multi-dimensional Schwarzian derivatives and differential operators

Lie algebras of vector fields on a smooth manifold M became popular in mathematics and physics after the discovery of the Virasoro algebra by Gelfand and Fuchs in 1967. Gelfand and Fuchs, Bott, Segal, Haefliger and many others studied cohomology of Lie algebras of vector fields and diffeomorphism groups with coefficients in spaces of tensor fields. This theory attracted much attention in the last three decades, many important problems were solved and many beautiful applications, such as characteristic classes of foliations, were found.

In this chapter we consider cohomology of Lie algebras of vector fields and of diffeomorphism groups with coefficients in various spaces of differential operators; this is a generalization of Gelfand–Fuchs cohomology. Only a few results are available so far, mostly for the first cohomology spaces. The main motivation is to study the space of differential operators $\mathcal{D}_{\lambda,\mu}(M)$, viewed as a module over the group of diffeomorphisms.

This cohomology is closely related to projective differential geometry and, in particular, to the Schwarzian derivative. The classic Schwarzian derivative is a 1-cocycle on the group $\mathrm{Diff}(S^1)$, related to the module of Sturm–Liouville operators. Multi-dimensional analogs of the Schwarzian derivative are defined as projectively invariant 1-cocycles on diffeomorphism groups with values in spaces of differential operators.

7.1 Multi-dimensional Schwarzian with coefficients in (2, 1)-tensors

In this section we consider a manifold M^n, $n \geq 2$, with a projective connection (see Appendix A.3) and describe the simplest version of multi-dimensional

Schwarzian derivative. This is a 1-cocycle of the group Diff(M) with values in symmetric $(2, 1)$-tensor fields on M. If the projective connection is flat, that is, M carries a projective structure, then this cocycle is projectively invariant. We also specify the definition in the case of a symplectic manifold M by restriction to the group of symplectomorphisms.

Space of $(2, 1)$-tensors

We start with some linear algebra. Let $V = \mathbb{R}^n$ be the standard $\mathrm{GL}(n, \mathbb{R})$-module. One has a natural projection

$$\mathrm{div} : S^2(V^*) \otimes V \to V^*;$$

if one identifies $S^2(V^*) \otimes V$ with component-wise quadratic vector fields then the projection indeed becomes the divergence operator. One also has a natural injection

$$j : V^* \to S^2(V^*) \otimes V,$$

defined by the formula

$$j(\ell) : (u, v) \mapsto \langle \ell, u \rangle v + u \langle \ell, v \rangle;$$

here $\ell \in V^*$ is a covector and $S^2(V^*) \otimes V$ is understood as the space of linear maps $S^2(V) \to V$.

Exercise 7.1.1 Check that $\mathrm{div} \circ j = (n + 1)\mathrm{Id}$.

Let $T_0 = \ker \mathrm{div}$. It follows that the formula

$$S \mapsto S - \frac{1}{n+1}(j \circ \mathrm{div})(S) \tag{7.1}$$

gives a projection $S^2(V^*) \otimes V \to T_0$. In fact,

$$S^2(V^*) \otimes V = T_0 \oplus \mathrm{im}\, j \tag{7.2}$$

is a decomposition on irreducible $\mathrm{GL}(n, \mathbb{R})$-modules.

Space of $(2, 1)$-tensor fields as a Vect(M)-module

Let $\mathcal{T}(M)$ be the space of symmetric 2-covariant 1-contravariant tensor fields on M; these fields are sections of the vector bundle $S^2(T^*M) \otimes TM$ over M. This space is naturally acted upon by the group Diff(M). Consider also the submodule $\mathcal{T}_0(M)$ of divergence free tensor fields of type T_0. The spaces $\mathcal{T}(M)$ and $\mathcal{T}_0(M)$ are also Vect(M)-modules.

Proposition 7.1.2 *If M is a closed manifold then*

$$H^1(\text{Vect}(M); \mathcal{T}_0(M)) = \mathbb{R}.$$

Proof The cohomology of the Lie algebra Vect(M) with coefficients in tensor fields was computed by Tsujishita, see [223] or [72]. The result of this computation is as follows:

$$H^*(\text{Vect}(M); \mathcal{A}) = H^*_{\text{top}}(Y(M)) \otimes \text{Inv}_{\text{gl(n},\mathbb{R})} \left(H^*(L_1; \mathbb{R}) \otimes A \right). \tag{7.3}$$

Here A is a gl($n, \mathbb{R}$)-module, $\mathcal{A}$ is the space of respective tensor fields, $Y(M)$ is a certain connected topological space, constructed from M (we do not need its description), and L_1 is the Lie algebra of formal vector fields in $\mathbb{R}^n$ with the trivial 1-jet.

In our case, $A = T_0$. Since we are interested in the first cohomology, we only need to consider the spaces

$$\text{Inv}_{\text{gl(n},\mathbb{R})} \left(H^0(L_1; \mathbb{R}) \otimes T_0 \right) \qquad \text{and} \qquad \text{Inv}_{\text{gl(n},\mathbb{R})} \left(H^1(L_1; \mathbb{R}) \otimes T_0 \right).$$

Since $H^0(L_1; \mathbb{R}) = 0$, the first of these spaces is trivial. To describe the second space, note that $H^1(\mathfrak{g}; \mathbb{R}) = (\mathfrak{g}/[\mathfrak{g}, \mathfrak{g}])^*$ for every Lie algebra $\mathfrak{g}$.

Exercise 7.1.3 Check that if $n \geq 2$ then $[L_1, L_1] = L_2$ where L_2 is the Lie algebra of formal vector fields in $\mathbb{R}^n$, with trivial 2-jet.

Hence $H^1(L_1; \mathbb{R}) = (L_1/L_2)^*$, the space of component-wise quadratic vector fields in $\mathbb{R}^n$, that is,

$$H^1(L_1; \mathbb{R}) = S^2(V) \otimes V^*.$$

It follows from (7.2) that

$$\text{Inv}_{\text{gl(n},\mathbb{R})} \left(H^1(L_1; \mathbb{R}) \otimes T_0 \right) = \mathbb{R},$$

and (7.3) implies that $H^1(\text{Vect}(M); \mathcal{T}_0(M))$ is one-dimensional. □

Introducing the multi-dimensional Schwarzian

The space of affine connections is an affine space with the underlining vector space $\mathcal{T}(M)$. This suggests using the "coboundaries of ghosts" trick described in Appendix A.5, namely, given an arbitrary affine connection ∇ on M, define a map $C : \mathrm{Diff}(M) \to \mathcal{T}(M)$ by

$$C(f) = \left(f^{-1}\right)^* \nabla - \nabla. \tag{7.4}$$

This map is clearly a 1-cocycle since it is written in the form of a "coboundary".

Lemma 7.1.4 *The cocycle C is non-trivial.*

Proof Any coboundary on $\mathrm{Diff}(M)$ with values in the space of tensor fields depends on 1-jets of diffeomorphisms. However, the transformation of connections involves second derivatives. □

Note that the same argument was used in Section 1.5 to prove non-triviality of the Schwarzian derivative.

Define a 1-cocycle

$$L : \mathrm{Diff}(M) \to \mathcal{T}_0(M) \tag{7.5}$$

by projecting the cocycle C from $\mathcal{T}(M)$ to $\mathcal{T}_0(M)$, see formula (7.1). The cocycle L is non-trivial for the same reason as C is.

Lemma 7.1.5 *The cocycle L depends only on the projective connection induced by ∇.*

Proof Two affine connections define the same projective connection if their difference has the trivial projection to $\mathcal{T}_0(M)$. □

The cocycle L will be called the *Schwarzian derivative with values in* (2, 1)-*tensor fields*. It is defined for an arbitrary manifold M of dimension ≥ 2 with a fixed projective connection. In the one-dimensional case, the cocycle L is identically zero.

Multi-dimensional Schwarzian and projective structures

Consider now the case $M = \mathbb{RP}^n$ and let L be the Schwarzian derivative associated with the canonical flat projective connection. As usual, we consider only differentiable cocycles (cf. Appendix A.4).

Theorem 7.1.6 *The cocycle L is the unique (up to a constant) non-trivial 1-cocycle on* $\mathrm{Diff}(\mathbb{RP}^n)$ *with values in* $\mathcal{T}_0(\mathbb{RP}^n)$*, vanishing on the subgroup* $\mathrm{PGL}(n+1, \mathbb{R})$.

Proof Let f be a diffeomorphism preserving the projective connection. Then $L(f) = 0$ by Lemma 7.1.5.

Let us prove the uniqueness. Consider two non-trivial 1-cocycles, L_1 and L_2, on $\mathrm{Diff}(\mathbb{RP}^n)$ with values in $\mathcal{T}_0(\mathbb{RP}^n)$. Let ℓ_1 and ℓ_2 be the respective 1-cocycles on $\mathrm{Vect}(\mathbb{RP}^n)$. They are cohomologous by Proposition 7.1.2: there is a linear combination which is a coboundary, namely $\alpha\ell_1 + \beta\ell_2 = d(t)$ where $t \in \mathcal{T}_0(\mathbb{RP}^n)$. If ℓ_1 and ℓ_2 vanish on $\mathrm{sl}(n+1, \mathbb{R})$, then t is projectively invariant.

Lemma 7.1.7 *There are no projectively invariant elements in $\mathcal{T}_0(\mathbb{RP}^n)$.*

Proof Let $t \in \mathcal{T}_0(\mathbb{RP}^n)$ be $\mathrm{PGL}(n+1, \mathbb{R})$-invariant. In an affine chart, t is invariant under the vector fields (6.5). The invariance with respect to translations implies that the coefficients of t are constants. But the space of tensors $\mathcal{T}_0$ is $\mathrm{GL}(n, \mathbb{R})$-irreducible. □

It follows that $\alpha\ell_1 + \beta\ell_2 = 0$. Finally, $\alpha L_1 + \beta L_2 = 0$ by Lemma 3.2.9. Theorem 7.1.6 is proved. □

If M is a manifold with a projective structure then a version of Theorem 7.1.6 holds: if $f \in \mathrm{Diff}(M)$ preserves the projective structure in an open domain U, then the tensor field $L(f)$ vanishes in U.

Expression in local coordinates

Fix a local coordinate system adapted to a projective structure on M.

Exercise 7.1.8 Check that, in local coordinates, the cocycle (7.5) is given by the formula

$$L(f) = \sum_{i,j,k} \left(\sum_{\ell} \frac{\partial^2 f^\ell}{\partial x^i \partial x^j} \frac{\partial x^k}{\partial f^\ell} - \frac{1}{n+1} \left(\delta^k_j \frac{\partial \log J_f}{\partial x^i} + \delta^k_i \frac{\partial \log J_f}{\partial x^j} \right) \right) \mathrm{d}x^i \otimes \mathrm{d}x^j \otimes \frac{\partial}{\partial x^k} \tag{7.6}$$

where $f(x^1, \ldots, x^n) = (f^1, \ldots, f^n)$ and $J_f = \det\left(\frac{\partial f^i}{\partial x^j}\right)$ is the Jacobian.

The first summand on the right hand side is the cocycle $C(f)$, one recognizes the standard expression of $\mathrm{Diff}(S^1)$-action on connections; the remaining terms come from the projection (7.2).

Cocycle L in the symplectic case

Consider the case when M is the symplectic plane $(\mathbb{R}^2, \mathrm{d}x \wedge \mathrm{d}y)$ and the group is $\mathrm{Diff}_0(\mathbb{R}^2)$, the group of area preserving diffeomorphisms.

The next geometric construction resembles that of the Schwarzian derivative in Section 1.3; one also may want to compare it with the material in Appendix A.5.

Given a symplectic diffeomorphism f, how can one measure its failure to be projective, that is, to send straight lines to straight lines? The following "triangle area construction" provides a natural answer. Let $x \in \mathbb{R}^2$ be a point and u a tangent vector at x. Consider the three collinear points $(x - \varepsilon u, x, x + \varepsilon u)$, where ε is a small parameter, and apply f to this triple. One obtains a triangle whose side lengths are of order ε and whose vertex angle is ε-close to π. It follows that the oriented area of the triangle is of order ε^3. Divide by ε^3, take the limit as $\varepsilon \to 0$, and denote the resulting number by $\bar{A}_{(x,u)}(f)$.

For a fixed f and x, by construction, $A_{(x,u)}(f)$ is a cubic form on the tangent space $T_x\mathbb{R}^2$. Hence A may be thought of as a map

$$A : \mathrm{Diff}_0(\mathbb{R}^2) \to \mathcal{C}(\mathbb{R}^2)$$

where $\mathcal{C}(\mathbb{R}^2)$ denotes the space of sections of $S^3(T^*\mathbb{R}^2)$, that is, the space of cubic forms on $\mathbb{R}^2$. This map depends on the 2-jet of a diffeomorphism and vanishes if the diffeomorphism is projective.

It is clear from the definition that

$$A(f \circ g) = g^* A(f) + A(g).$$

It follows that the map $f \mapsto A(f^{-1})$ is a 1-cocycle on the group of symplectomorphisms with coefficients in cubic differentials.

Exercise 7.1.9 Prove the explicit formula:

$$A(f) = \begin{vmatrix} \phi_x & \psi_x \\ \phi_{xx} & \psi_{xx} \end{vmatrix} \mathrm{d}x^3 + 3 \begin{vmatrix} \phi_x & \psi_x \\ \phi_{xy} & \psi_{xy} \end{vmatrix} \mathrm{d}x^2\mathrm{d}y$$

$$+ 3 \begin{vmatrix} \phi_y & \psi_y \\ \phi_{xy} & \psi_{xy} \end{vmatrix} \mathrm{d}x\mathrm{d}y^2 + \begin{vmatrix} \phi_y & \psi_y \\ \phi_{yy} & \psi_{yy} \end{vmatrix} \mathrm{d}y^3$$

where $(\phi(x, y), \psi(x, y))$ are the components of the symplectomorphism f and where ϕ_x stands for a partial derivative.

The corresponding 1-cocycle of the Lie algebra of Hamiltonian vector fields is:

$$a(X_F) = F_{xxx}\, \mathrm{d}x^3 + 3F_{xxy}\, \mathrm{d}x^2\mathrm{d}y + 3F_{xyy}\, \mathrm{d}x\mathrm{d}y^2 + F_{yyy}\, \mathrm{d}y^3$$

where X_F is the symplectic gradient of a function $F(x, y)$.

The triangle area construction extends verbatim to the case when f is a symplectic diffeomorphism of $\mathbb{R}^{2n}$: one uses the symplectic form to measure areas of triangles.

The relation between the cocycle A and the restriction of the cocycle L to the group $\mathrm{Diff}_0(\mathbb{R}^{2n})$ is as follows. The isomorphism $V^* = V$ for the linear symplectic space, $V = \mathbb{R}^{2n}$, along with the natural projection $S^2V^* \otimes V^* \to S^3V^*$, provides the map

$$\pi : S^2V^* \otimes V = S^2V^* \otimes V^* \to S^3V^*.$$

Denote by the same symbol the map of the spaces of sections.

Exercise 7.1.10 Check that $A = \pi \circ L$.

Comment

The usefulness of connections in constructing non-trivial cocycles on $\mathrm{Diff}(M)$ (and $\mathrm{Vect}(M)$) was emphasized by Koszul [125]. A different proof of Proposition 7.1.2 can be found in [96]. Another result of this paper is $H^1(\mathrm{Vect}(M); \mathcal{T}(M)) = \mathbb{R}^2$.

The multi-dimensional Schwarzian derivative introduced in this section has been around for quite a while (see, e.g., [73, 124, 234, 155]); most of these references deal with the complex analytic case. Theorem 7.1.6 is new.

The case of symplectic manifolds was considered in [202], the triangle area construction was suggested by E. Ghys. The symplectic version of the multi-dimensional Schwarzian derivative gives rise to a 2-cocycle on $\mathrm{Diff}_0(M)$ with coefficients in functions on a symplectic manifold M, the so-called group Vey cocycle; see [202].

7.2 Projectively equivariant symbol calculus in any dimension

In this section we consider the space, $\mathcal{D}_{\lambda,\mu}(M)$, of differential operators on a manifold M

$$A : \mathcal{F}_\lambda(M) \to \mathcal{F}_\mu(M)$$

where $\mathcal{F}_\lambda(M)$ and $\mathcal{F}_\mu(M)$ are the spaces of tensor densities of degree λ and μ. In the one-dimensional case, this space was thoroughly studied in Chapter 3, and played an important role in the theory of projective curves.

As in the one-dimensional case, we study $\mathcal{D}_{\lambda,\mu}(M)$ as a module over the group of diffeomorphisms $\mathrm{Diff}(M)$ and the Lie algebra of vector fields $\mathrm{Vect}(M)$.

We will be particularly interested in the case when M is equipped with a projective structure. We then construct a canonical $\mathrm{sl}(n+1, \mathbb{R})$-isomorphism between the space $\mathcal{D}_{\lambda,\mu}(M)$ and the corresponding space of symbols.

Space of differential operators and space of symbols

The space of differential operators $\mathcal{D}_{\lambda,\mu}(M)$ is a filtered Diff(M)-module:

$$\mathcal{D}^0_{\lambda,\mu}(M) \subset \mathcal{D}^1_{\lambda,\mu}(M) \subset \cdots \subset \mathcal{D}^k_{\lambda,\mu}(M) \subset \cdots$$

where $\mathcal{D}^k_{\lambda,\mu}(M)$ is the space of operators of order k. The associated graded module $\mathrm{gr}\left(\mathcal{D}_{\lambda,\mu}(M)\right)$ is called the *space of symbols*.

Example 7.2.1 If $\lambda = \mu$, then the space of symbols has a simple geometric interpretation. It is naturally isomorphic (i.e., isomorphic as a Diff(M)-module) to the space of symmetric contravariant tensor fields on M. This space will be denoted by $\mathcal{S}(M)$, it also can be viewed as the space of fiber-wise polynomial functions on T^*M.

More generally, there is an isomorphism of Diff(M)-modules

$$\mathrm{gr}\left(\mathcal{D}_{\lambda,\mu}(M)\right) \cong \mathcal{S}(M) \otimes_{C^\infty(M)} \mathcal{F}_\delta(M)$$

where $\delta = \mu - \lambda$. We will use the notation $\mathcal{S}_\delta(M)$ for the above module of symbols. This is a graded Diff(M)-module:

$$\mathcal{S}_\delta(M) = \bigoplus_{k=0}^{\infty} \mathcal{S}_{k,\delta}(M)$$

where $\mathcal{S}_{k,\delta}(M)$ corresponds to the space of polynomials of degree k.

Formulæ in local coordinates

In local coordinates, a tensor density of degree λ is written as follows:

$$\varphi = f(x^1, \dots, x^n)\left(\mathrm{d}x^1 \wedge \cdots \wedge \mathrm{d}x^n\right)^\lambda$$

where f is a smooth function. A differential operator $A \in \mathcal{D}_{\lambda,\mu}(M)$ of order k is of the form

$$A(\varphi) = \Big(\sum_{i_1,\dots,i_k} A_k^{i_1,\dots,i_k} \frac{\partial^k f}{\partial x^{i_1} \cdots \partial x^{i_k}} + \cdots + A_0 f\Big)\left(\mathrm{d}x^1 \wedge \cdots \wedge \mathrm{d}x^n\right)^\mu.$$

A symbol of degree k can be written as

$$P = \sum_{i_1,\ldots,i_k} P_k^{i_1,\ldots,i_k}\, \xi_{i_1}\cdots\xi_{i_k} + \cdots + P_0$$

where $\xi_1,\ldots,\xi_n$ are the Darboux coordinates on the fibers of T^*M.

Statement of the problem

Assume now that manifold M is equipped with a projective structure. This means that there is a (locally defined) action of the Lie group $\mathrm{PGL}(n+1,\mathbb{R})$ and the Lie algebra $\mathrm{sl}(n+1,\mathbb{R})$ on M; see Section 6.1.

We are looking for a "total" *symbol map*

$$\sigma_{\lambda,\mu} : \mathcal{D}_{\lambda,\mu}(M) \longrightarrow \mathcal{S}_\delta(M)$$

that commutes with the $\mathrm{PGL}(n+1,\mathbb{R})$-action. In other words, we want to identify these two spaces canonically with respect to the projective structure on M. The inverse of the symbol map,

$$\mathcal{Q}_{\lambda,\mu} = \left(\sigma_{\lambda,\mu}\right)^{-1},$$

is called the *quantization map*. A symbol map and a quantization map, commuting with the $\mathrm{PGL}(n+1,\mathbb{R})$-action, are called *projectively equivariant*. We want to give a complete classification of such maps.

In the one-dimensional case, this problem was solved in Section 2.5.

Commutant of the affine Lie algebra

Fix a system of local coordinates on M adapted to the projective structure. The action of the Lie algebra $\mathrm{sl}(n+1,\mathbb{R})$ is given by formula (6.5).

Consider the action of the affine subalgebra $\mathrm{aff}(n,\mathbb{R})$ on the space of differential operators.

Exercise 7.2.2 Show that the actions of $\mathrm{aff}(n,\mathbb{R})$ on $\mathcal{D}_{\lambda,\mu}(M)$ and on $\mathcal{S}_\delta(M)$, written in the coordinates of the projective structure, are identically the same.

Therefore one locally identifies these two spaces as $\mathrm{aff}(n,\mathbb{R})$-modules.

Consider the differential operators

$$\mathcal{E} = \sum_{i=1}^n \xi_i \frac{\partial}{\partial \xi_i}, \qquad \mathrm{D} = \sum_{i=1}^n \frac{\partial}{\partial x^i}\frac{\partial}{\partial \xi_i}, \tag{7.7}$$

defined (locally) on T^*M. The first one is called the *Euler operator* and the second the *divergence operator*.

To simplify the notation, consider the action of the operators $\mathcal{E}$ and D on the space of polynomials $\mathbb{C}[x^1, \ldots, x^n, \xi_1, \ldots, \xi_n]$. The next statement is the classic Weyl–Brauer theorem, well-known in invariant theory; see [85]: *The commutant of the* aff$(n, \mathbb{R})$*-action on* $\mathbb{C}[x^1, \ldots, x^n, \xi_1, \ldots, \xi_n]$ *is the associative algebra with generators* $\mathcal{E}$ *and* D.

A projectively equivariant symbol map $\sigma_{\lambda,\mu}$ has to commute with the aff$(n, \mathbb{R})$-action as well. It follows that a projectively equivariant symbol and quantization maps, in coordinates adapted to the projective structure, have to be expressions in the operators $\mathcal{E}$ and D.

The main result

As in the one-dimensional case (cf. Section 2.5), the PGL$(n+1, \mathbb{R})$-invariant symbol map is given by a confluent hypergeometric function (2.23) with parameters depending on $\mathcal{E}$ and the argument D.

Theorem 7.2.3 *The* PGL$(n+1, \mathbb{R})$*-modules* $\mathcal{D}_{\lambda,\mu}(M)$ *and* $\mathcal{S}_\delta(M)$ *are isomorphic, provided*

$$\delta \neq 1 + \frac{\ell}{n+1}. \tag{7.8}$$

The projectively equivariant symbol map is unique (up to a constant) and given by

$$\sigma_{\lambda,\mu} = F\left(\begin{matrix} a \\ b \end{matrix} \middle| z\right), \tag{7.9}$$

where

$$a = \mathcal{E} + (n+1)\lambda, \qquad b = 2\mathcal{E} + (n+1)(1-\delta), \qquad z = -\mathrm{D}. \tag{7.10}$$

Proof Consider the (locally defined) map (7.9) in a coordinate system adapted to the projective structure on M. Let us rewrite this map, restricted to each component $\mathcal{S}_{k,\delta}(M)$ of a fixed order k. We will then reformulate the theorem in this special case. The most important simplification is

$$\mathcal{E}|_{\mathcal{S}_{k,\delta}(M)} = k\,\mathrm{Id},$$

so that the expression for the map $\sigma_{\lambda,\mu}$ in homogeneous k-th order polynomials will contain only the operator D.

Exercise 7.2.4 Check that the map (7.9), restricted to the k-th order component, is given by

$$\sigma_{\lambda,\mu}\Big|_{\mathcal{S}_{k,\delta}(M)} = \sum_{\ell=O}^{k} C_\ell^k \frac{\mathrm{D}^\ell}{\ell!}, \tag{7.11}$$

where

$$C_\ell^k = (-1)^\ell \frac{\binom{(n+1)\lambda+k-1}{\ell}}{\binom{2k-\ell+(n+1)(1-\delta)-1}{\ell}}; \tag{7.12}$$

here $\binom{a}{b}$ is the binomial coefficient.

Formula (7.11)–(7.12) is equivalent to (7.9)–(7.10), and it makes sense if δ satisfies (7.8). It then suffices to prove that a $\mathrm{PGL}(n+1,\mathbb{R})$-invariant symbol map is given by this expression.

Exercise 7.2.5 Consider a map on $\mathcal{S}_{k,\delta}(M)$ given by formula (7.11) with undetermined coefficients C_ℓ^k. Imposing the equivariance condition with respect to the quadratic vector fields $X_i = x^i \sum_j x^j \partial/\partial x^j$, prove that these coefficients are as in (7.12).

We proved that the symbol map $\sigma_{\lambda,\mu}$, defined (locally on M) by (7.11)–(7.12), is $\mathrm{PGL}(n+1,\mathbb{R})$-invariant. Therefore, this map does not change under the linear-fractional coordinate transformations (6.1) and is globally defined. This completes the proof. □

Example of second-order differential operators

Consider the space $\mathcal{D}_\lambda^2(M) = \mathcal{D}_{\lambda,\lambda}^2(M)$ of second-order differential operators in the particular case $\mu = \lambda$. In local coordinates, a second-order differential operator is given by

$$A = \sum_{i,j=1}^{n} A_2^{ij} \frac{\partial}{\partial x^i}\frac{\partial}{\partial x^j} + \sum_{i=1}^{n} A_1^i \frac{\partial}{\partial x^i} + A_0 \tag{7.13}$$

where the coefficients A_2^{ij}, A_1^i and A_0 are smooth functions in x. The explicit formula for the projectively equivariant symbol map

$$\sigma_\lambda : \mathcal{D}_\lambda^2(M) \to \mathcal{S}_2(M) \oplus \mathcal{S}_1(M) \oplus \mathcal{S}_0(M)$$

is, of course, a particular case of (7.9)–(7.10).

Exercise 7.2.6 Check that the map σ_λ associates with a differential operator (7.13) the second-order polynomial

$$P = \sum_{i,j=1}^{n} P_2^{ij}\,\xi_i\xi_j + \sum_{i=1}^{n} P_1^i\,\xi_i + P_0 \tag{7.14}$$

given, in a coordinate system adapted to the projective structure, by

$$\begin{aligned} P_2^{ij} &= A_2^{ij} \\ P_1^i &= A_1^i - 2\frac{(n+1)\lambda+1}{n+3}\sum_{j=1}^{n}\frac{\partial A_2^{ij}}{\partial x^j} \\ P_0 &= A_0 - \lambda\sum_{i=1}^{n}\frac{\partial A_1^i}{\partial x^i} + \lambda\frac{(n+1)\lambda+1}{n+2}\sum_{i,j=1}^{n}\frac{\partial^2 A_2^{ij}}{\partial x^i\partial x^j}. \end{aligned} \tag{7.15}$$

The quantization map

One can also write an explicit formula for the quantization map. Let us give it in the most interesting particular case $\lambda = \mu = 1/2$ when this map is simpler. We will use the notation $\mathrm{E} = \mathcal{E} + \frac{1}{2}n$. One has

$$\mathcal{Q}_{1/2,1/2} = F\left(\begin{array}{c|c} 2\mathrm{E} & \frac{\mathrm{D}}{4} \\ \mathrm{E} & \end{array}\right) \tag{7.16}$$

The quantization map (7.16) has nice properties. The following statement is a consequence of the uniqueness of the symbol map (and thus of the quantization map).

Corollary 7.2.7 *Substitute* $i\,\mathrm{D}$ *instead of* D *in operator (7.16). Then, for every real* $P \in \mathcal{S}_\delta(M)$, *the "complexified" differential operator* $\mathcal{Q}_{1/2,1/2}(P)$ *is symmetric.*

Proof The symmetrized map

$$P \mapsto \frac{1}{2}\left(\mathcal{Q}_{1/2,1/2}(P) + \mathcal{Q}_{1/2,1/2}(P)^*\right)$$

is also $\mathrm{PGL}(n+1,\mathbb{R})$-invariant. It has the same principal symbol. The uniqueness part of Theorem 7.2.3 implies that this map coincides with $\mathcal{Q}_{1/2,1/2}$, and the result follows. □

Comment

The subject of this section belongs to a research program *equivariant quantization* based on the general equivariance principle: the symbol and quantization maps should commute with the action of a Lie group on M. There are few examples for which the equivariance condition uniquely determines the symbol map. Projective differential geometry is the first example, another is conformal differential geometry.

The projectively equivariant symbol and quantization maps in the multi-dimensional case were introduced in [134]. The expression in terms of hypergeometric functions was suggested in [57]. In the one-dimensional case, the projectively equivariant symbol and quantization maps were found in [44] (the results of [44] and [134] are independent). The conformal case was treated in [54].

The special values δ from (7.8) are called the *resonant* values. For every such δ, there are particular values of λ for which the isomorphism $\sigma_{\lambda,\mu}$ still exists. The resonant $\mathrm{PGL}(n+1,\mathbb{R})$-modules of differential operators were classified in [131], see also [132].

7.3 Multi-dimensional Schwarzian as a differential operator

Let M be a manifold equipped with a projective structure. We study the space $\mathcal{D}^2_\lambda(M)$ of second-order differential operators on the space of tensor densities $\mathcal{F}_\lambda(M)$. As a $\mathrm{PGL}(n+1,\mathbb{R})$-module, this space is canonically isomorphic to the corresponding space of symbols. The multi-dimensional analog of the Schwarzian derivative we introduce in this section measures the failure of the two modules to be isomorphic as $\mathrm{Diff}(M)$-modules. This is a 1-cocycle on $\mathrm{Diff}(M)$ vanishing on $\mathrm{PGL}(n+1,\mathbb{R})$. Unlike the multi-dimensional Schwarzian derivative of Section 7.1, it depends on the 3-jet of a diffeomorphism. Moreover, this 1-cocycle on $\mathrm{Diff}(M)$ coincides with the classic Schwarzian derivative in the case $\dim M = 1$.

The ideas are similar to those of Sections 3.2 and 3.3.

Formulating the problem

The space $\mathcal{D}^2_\lambda(M)$ is a $\mathrm{Diff}(M)$-module with respect to the action $T^{\lambda,\lambda}$, defined by formula (2.1) with $\mu = \lambda$. Our goal is to compare this space and the corresponding space of symbols as modules over $\mathrm{Diff}(M)$.

The space of symbols is the space of fiber-wise polynomial functions on T^*M of order ≤ 2. As a module over $\mathrm{Diff}(M)$, it splits into a direct sum

$$\mathcal{S}^2(M) = \mathcal{S}_2(M) \oplus \mathrm{Vect}(M) \oplus C^\infty(M)$$

where $\mathcal{S}_2(M)$ is the space of homogeneous second-order polynomials on T^*M, or, equivalently, of symmetric contravariant tensor fields on M, namely, $\mathcal{S}_2(M) = \Gamma(S^2TM)$. An element P of $\mathcal{S}^2(M)$ is of the form (7.14).

The spaces $\mathcal{D}^2_\lambda(M)$ and $\mathcal{S}^2(M)$ are isomorphic as $\mathrm{PGL}(n+1, \mathbb{R})$-modules (see Theorem 7.2.3); the isomorphism is defined by formula (7.15). As in Section 3.3, let us rewrite the Diff(M)-action on $\mathcal{D}^2_\lambda(M)$ in a projectively invariant way. More precisely, we consider the Diff(M)-action

$$f(P) = \sigma_\lambda \circ \mathrm{T}_f^{\lambda,\lambda} \circ \sigma_\lambda^{-1}(P) \tag{7.17}$$

which is the usual action on $\mathcal{D}^2_\lambda(M)$, written in terms of $\mathcal{S}^2(M)$.

Appearance of 1-cocycles

Let us compute the above defined action explicitly.

Exercise 7.3.1 Check that the action (7.17) is as follows:

$$\begin{aligned} f(P)_2 &= f_* P_2, \\ f(P)_1 &= f_* P_1 \ + \ \frac{n+1}{n+3}\,(2\lambda - 1)\,\mathcal{L}(f)\,(f_* P_2), \\ f(P)_0 &= f_* P_0 \ - \ \frac{2}{n+2}\,\lambda(\lambda - 1)\,\mathcal{S}(f)\,(f_* P_2), \end{aligned} \tag{7.18}$$

where f_* is the natural action of f on the tensor fields, while $\mathcal{L}(f)$ and $\mathcal{S}(f)$ are linear differential operators

$$\mathcal{L}(f) : \mathcal{S}_2(M) \to \mathrm{Vect}(M), \qquad \mathcal{S}(f) : \mathcal{S}_2(M) \to C^\infty(M).$$

The fact that formula (7.18) indeed defines an action of Diff(M) already implies that the maps $\mathcal{L}$ and $\mathcal{S}$ are 1-cocycles on Diff(M):

$$\mathcal{L} \in Z^1(\mathrm{Diff}(M); \mathcal{D}(\mathcal{S}_2(M), \mathrm{Vect}(M))),$$

$$\mathcal{S} \in Z^1(\mathrm{Diff}(M); \mathcal{D}(\mathcal{S}_2(M), C^\infty(M))).$$

By construction, these cocycles are projectively invariant. Our next task is to compute the above 1-cocycles explicitly.

Computing the 1-cocycles

The cocycle $\mathcal{L}$ is nothing else but the multi-dimensional Schwarzian derivative introduced in Section 7.1, viewed as a zero-order differential operator.

Exercise 7.3.2 The 1-cocycle $\mathcal{L}$ is given by the contraction with the cocycle (7.5), namely $\mathcal{L}(f)(P) := \langle L(f), P\rangle$.

In local coordinates of the projective structure, one has

$$\mathcal{L}(f)(P)^i = \sum_{j,k=1}^{n} L(f)^i_{jk}\, P^{jk}$$

where the coordinate expression for L is given by formula (7.6).

The 1-cocycle $\mathcal{S}$ is more complicated. It is given by a first-order differential operator from $\mathcal{S}_2(M)$ to $C^\infty(M)$. Let us compute its explicit coordinate formula.

Exercise 7.3.3 Check that in local coordinates, adapted to the projective structure, one has

$$\mathcal{S}(f)(P) = \sum_{ij} S(f)_{ij}(P^{ij})$$

where

$$S(f)_{ij} = \sum_{k=1}^{n} L(f)^k_{ij}\,\frac{\partial}{\partial x^k}$$
$$-\frac{2}{n-1}\sum_{k=1}^{n}\frac{\partial}{\partial x^k}\left(L(f)^k_{ij}\right) + \frac{n+1}{n-1}\sum_{k,\ell=1}^{n} L(f)^k_{i\ell}\, L(f)^\ell_{kj}. \quad (7.19)$$

The map $\mathcal{S}$, defined by formula (7.19), clearly does not depend on the choice of a coordinate system, adapted to the projective structure.

Uniqueness

As in Section 7.1, we will prove the uniqueness of the multi-dimensional Schwarzian derivatives $\mathcal{L}$ and $\mathcal{S}$. The main result of this section is as follows.

Theorem 7.3.4 *The cocycles $\mathcal{L}$ and $\mathcal{S}$ are the unique, up to a multiplicative constant, projectively invariant non-trivial 1-cocycles on* Diff(M) *with values in $\mathcal{D}(\mathcal{S}_2(M), \mathrm{Vect}(M))$ and $\mathcal{D}(\mathcal{S}_2(M), C^\infty(M))$, respectively.*

Proof Let us first prove that the 1-cocycles $\mathcal{L}$ and $\mathcal{S}$ are non-trivial. The cocycle $\mathcal{S}$ depends on the third jet of f. However, every coboundary is of the form $f^*(B) - B$ where $B \in \mathcal{D}(\mathcal{S}_2(M), C^\infty(M))$. Since $S(f)$ is a first-order differential operator for all f, the coboundary condition $S = d(B)$ would imply that B is also a first-order differential operator and thus $d(B)(f)$ depends on at most the second jet of f. The proof of non-triviality of $\mathcal{L}$ is similar.

Let us prove the uniqueness. The main ingredient of the proof is again a cohomological statement.

Proposition 7.3.5 *The following two cohomology spaces are one-dimensional for every closed manifold M:*

$$H^1(\mathrm{Diff}(M); \mathcal{D}(\mathcal{S}_2(M), \mathrm{Vect}(M))) = \mathbb{R},$$

$$H^1(\mathrm{Diff}(M); \mathcal{D}(\mathcal{S}_2(M), C^\infty(M))) = \mathbb{R}.$$

The proof of this statement will be omitted; see [133].

Given two projectively invariant cocycles C_1 and C_2, there is a linear combination $\alpha C_1 + \beta C_2$ which is a coboundary, $d(B)$. The differential operator B is then projectively invariant. It follows that the principal symbol of B is also projectively invariant. However, there are no non-zero tensor fields on M which are projectively invariant. Therefore $B = 0$, and so C_1 and C_2 are proportional. Theorem 7.3.4 is proved. □

Comment

The projectively invariant cocycle (7.19) was introduced in [34] in the projectively flat case and generalized for arbitrary projective connections in [32]. Cohomology of $\mathrm{Diff}(M)$ with coefficients in the spaces of differential operators $\mathcal{D}(\mathcal{S}_2(M), \mathrm{Vect}(M))$ and $\mathcal{D}(\mathcal{S}_2(M), C^\infty(M))$ were computed in [133]. Theorem 7.3.4 on the uniqueness of the cocycles $\mathcal{L}$ and $\mathcal{S}$ is new.

7.4 Application: classification of modules $\mathcal{D}^2_\lambda(M)$ for an arbitrary manifold

Let M be an arbitrary smooth manifold with $\dim M \geq 2$. We will classify the modules $\mathcal{D}^2_\lambda(M)$. In other words, we determine all values of λ and λ' for which there is an isomorphism

$$\mathcal{D}^2_\lambda(M) \cong \mathcal{D}^2_{\lambda'}(M).$$

Surprisingly enough, the results on projectively invariant symbol calculus and multi-dimensional analogs of the Schwarzian derivative (see Sections 7.2 and 7.3) can be applied even in the case when M is not equipped with a projective structure.

Classification theorem

The main result of this section is the following theorem.

Theorem 7.4.1 *The modules $\mathcal{D}^2_\lambda(M)$ are isomorphic to each other, provided $\lambda \neq 0, \frac{1}{2}, 1$. The two modules $\mathcal{D}^2_{\frac{1}{2}}(M)$ and $\mathcal{D}^2_0(M) \cong \mathcal{D}^2_1(M)$ are exceptional and not isomorphic to each other or to $\mathcal{D}^2_\lambda(M)$ with a generic λ.*

Proof Fix an arbitrary system of local coordinates $(x^1, \dots, x^n)$ in an open domain $U \subset M$ and consider the locally defined symbol (7.15) in these coordinates. Consider the linear map $\mathcal{I}_{\lambda,\lambda'} : \mathcal{D}^2_\lambda(U) \to \mathcal{D}^2_{\lambda'}(U)$ defined, in terms of the projectively equivariant symbol, by

$$\mathcal{I}_{\lambda,\lambda'} : P_2 + P_1 + P_0 \longmapsto P_2 + \frac{2\lambda - 1}{2\lambda' - 1} P_1 + \frac{\lambda(\lambda - 1)}{\lambda'(\lambda' - 1)} P_0 \tag{7.20}$$

This map intertwines two actions (7.18) with λ and λ', provided λ and λ' are different from 0, $\frac{1}{2}$ and 1.

The crucial point of the proof is that the map $\mathcal{I}_{\lambda,\lambda'}$ is *globally defined* on M. In other words, formula (7.20) does not depend on the choice of coordinates. This is equivalent to the fact that $\mathcal{I}_{\lambda,\lambda'}$ commutes with the Diff(M)-action (7.18).

Exercise 7.4.2 Check this.

This map is an isomorphism between $\mathcal{D}^2_\lambda(M)$ and $\mathcal{D}^2_{\lambda'}(M)$.

The modules corresponding to the values 0, $\frac{1}{2}$ and 1 of the parameter λ are not isomorphic to the generic ones. This follows from the fact that the 1-cocycles L and S are non-trivial. Finally, the modules $\mathcal{D}^2_0(M)$ and $\mathcal{D}^2_1(M)$ are isomorphic (by conjugation) but not isomorphic to $\mathcal{D}^2_{\frac{1}{2}}(M)$ since these modules correspond to different cohomology classes. □

The isomorphism $\mathcal{I}_{\lambda,\lambda'}$

The isomorphism from Theorem 7.4.1 enjoys quite remarkable properties.

Proposition 7.4.3 *(i) Isomorphism between the* Diff(M)*-modules $\mathcal{D}^2_\lambda(M)$ and $\mathcal{D}^2_{\lambda'}(M)$ is unique up to a constant and is given by (7.20).*
(ii) If $\lambda = \lambda'$, then the map (7.20) is the identity; if $\lambda + \lambda' = 1$, then it coincides with the conjugation of differential operators.

Proof Locally, over the domain U, one can restrict the Diff(M)-module structure to PGL$(n+1, \mathbb{R})$. Then the isomorphism $\mathcal{I}_{\lambda,\lambda'}$ has to commute with the PGL$(n+1, \mathbb{R})$-action. The uniqueness then follows from the uniqueness of the projectively equivariant symbol map, see Theorem 7.2.3.

Part (ii) is easily seen from formula (7.20). □

Let us now give the formula for the isomorphism $\mathcal{I}_{\lambda,\lambda'}$ directly in terms of the coefficients of differential operators.

Exercise 7.4.4 Check that the map (7.20) associates with a differential operator (7.13) a differential operator $\widetilde{A} = \mathcal{I}_{\lambda,\lambda'}(A)$ with the coefficients

$$\widetilde{A}_2^{ij} = A_2^{ij},$$

$$\widetilde{A}_1^i = \frac{2\lambda' - 1}{2\lambda - 1} A_1^i + 2\frac{\lambda' - \lambda}{2\lambda - 1} \sum_{j=1}^n \frac{\partial A_2^{ij}}{\partial x^j},$$

$$\widetilde{A}_0 = \frac{\lambda'(\lambda' - 1)}{\lambda(\lambda - 1)} A_0 - \frac{\lambda'(\lambda' - \lambda)}{(2\lambda - 1)(\lambda - 1)} \left(\sum_{i=1}^n \frac{\partial A_1^i}{\partial x^i} - \sum_{i,j=1}^n \frac{\partial^2 A_2^{ij}}{\partial x^i \partial x^j} \right).$$

It is hard to believe that the above expression is, actually, invariantly defined. This follows, nevertheless, from the fact that the map $\mathcal{I}_{\lambda,\lambda'}$ commutes with the Diff(M)-action. The isomorphism $\mathcal{I}_{\lambda,\lambda'}$ is an example of an invariant (or natural) differential operator (see Comment in Section 2.1).

Exercise 7.4.5 Express the map $\mathcal{I}_{\lambda,\lambda'}$ in terms of the Lie derivative:

$$\mathcal{I}_{\lambda,\lambda'}\left(L_X^\lambda \circ L_Y^\lambda\right) = L_X^{\lambda'} \circ L_Y^{\lambda'} + \frac{1}{\lambda + \lambda'}\left[L_X^{\lambda'}, L_Y^{\lambda'}\right],$$

$$\mathcal{I}_{\lambda,\lambda'}\left(L_X^\lambda\right) = \frac{2\lambda - 1}{2\lambda' - 1} L_X^{\lambda'},$$

$$\mathcal{I}_{\lambda,\lambda'}(F) = \frac{\lambda(\lambda - 1)}{\lambda'(\lambda' - 1)} F,$$

where X, Y are vector fields and F is a function.

Written in this form, the map $\mathcal{I}_{\lambda,\lambda'}$ is obviously coordinate free.

Comment

Theorem 7.4.1 was proved in [55]. The isomorphism $\mathcal{I}_{\lambda,\lambda'}$ was also introduced in this paper and called there the "second-order Lie derivative" since it defines an action of a differential operator on tensor densities of different degrees. The modules $\mathcal{D}^k_{\lambda,\mu}(M)$ were recently classified for arbitrary k and any values of λ and μ; see [143] and references therein.

7.5 Poisson algebra of tensor densities on a contact manifold

Projective differential geometry of $\mathbb{RP}^1$ was thoroughly studied in the first chapters of this book. We saw that the one-dimensional case is particularly rich: in addition to various geometric notions, complicated algebraic structures, such as the algebra of differential operators, the Poisson algebra of tensor densities and the Virasoro algebra, naturally appeared in the context. One would expect the situation to become even more intriguing in the multi-dimensional case. However, this expectation is not fulfilled: the Lie algebra of all vector fields on a smooth manifold of dimension greater than 1 has no central extensions. There is no multi-dimensional analog of the Virasoro algebra in this sense, and the space of tensor densities has no natural Poisson structure.

Recent developments in symplectic and contact topology suggest another viewpoint that can be expressed by the following "proportion":

$$\frac{\text{affine}}{\text{projective}} \simeq \frac{\text{symplectic}}{\text{contact}}.$$

One is then led to consider the odd-dimensional projective space $\mathbb{RP}^{2n-1}$, equipped with the canonical contact structure, as a multi-dimensional analog of $\mathbb{RP}^1$. Of course, the symmetry group becomes smaller, namely, $\mathrm{Sp}(2n, \mathbb{R})$ instead of $\mathrm{PGL}(2n, \mathbb{R})$, while the set of invariants gets richer.

The space of tensor densities on a contact manifold is a Poisson algebra. We consider the Poisson algebra of tensor densities, $\mathcal{F}(M)$, where M is a contact manifolds equipped with a compatible projective structure. For example, $M = \mathbb{RP}^{2n-1}$ or S^{2n-1} with the canonical contact and projective structures. There are natural analogs of transvectants and a star-product on $\mathcal{F}(M)$ and a series of extensions of the Lie algebra of contact vector fields on M. The situation is quite similar to that in the one-dimensional case; see Sections 3.4 and 3.5.

A particularly interesting case is that of $\dim M = 4k + 1$ when one obtains Lie algebras that have non-trivial central extensions. These algebras are considered as multi-dimensional analogs of the Virasoro algebra.

Symplectization

Let M^{2n-1} be a contact manifold; see Appendix A.2. Consider the $2n$-dimensional submanifold S of the cotangent bundle T^*M that consists of all covectors vanishing on the contact distribution ξ. The following statement is classic; see [15].

Proposition 7.5.1 *The restriction to S of the canonical symplectic structure on T^*M defines a symplectic structure on S.*

The manifold S is called the *symplectization* of M. Clearly S is a line bundle over M; its sections are the 1-forms on M vanishing on ξ.

Let $\mathrm{Diff}_c(M)$ be the group of *contact diffeomorphisms* of M, that is, of the diffeomorphisms of M that preserve the contact structure. This group naturally acts on the bundle S and therefore on the space of sections $\mathrm{Sec}(S)$. Let us prove that sections of the bundle S can be viewed as tensor densities of degree $1/n$ on M.

Proposition 7.5.2 *There is a natural isomorphism of* $\mathrm{Diff}_c(M)$*-modules*

$$\mathrm{Sec}(S) = \mathcal{F}_{1/n}(M). \tag{7.21}$$

Proof Let α be a contact form. Then $\Omega = \alpha \wedge (\mathrm{d}\alpha)^{n-1}$ is a volume form. If f is a contact diffeomorphism, then $f^*\alpha$ is proportional to α:

$$f^*\alpha = m_f\,\alpha$$

where m_f is a non-vanishing function. One then has $f^*\Omega = (m_f)^n\,\Omega$ so that the contact form α, indeed, transforms as a tensor density of degree $1/n$. □

Poisson algebra of tensor densities

Let us define a structure of Poisson algebra on the space $\mathcal{F}(M)$ of tensor densities on M. The definition is quite similar to that in the one-dimensional case; see Section 3.4. We identify tensor densities on M with homogeneous functions on $S \setminus M$, the symplectization with the zero section removed, and then use the Poisson bracket on S.

The following statement immediately follows from Proposition 7.5.2.

Corollary 7.5.3 *The space* $\mathcal{F}_\lambda(M)$ *is isomorphic, as a module over the group* $\mathrm{Diff}_c(M)$*, to the space of homogeneous functions on* $S \setminus M$ *of degree* $-\lambda\,n$.

The Poisson bracket on S is homogeneous of degree -1, so that the bracket of two homogeneous functions of degree $-\lambda\,n$ and $-\mu\,n$ is a homogeneous function of degree $-\lambda\,n - \mu\,n - 1$. One then obtains a bilinear differential operator on tensor densities on M:

$$\{\,,\,\} : \mathcal{F}_\lambda(M) \otimes \mathcal{F}_\mu(M) \to \mathcal{F}_{\lambda+\mu+\frac{1}{n}}(M), \tag{7.22}$$

so that the space $\mathcal{F}(M)$ is equipped with a natural structure of a Poisson algebra.

Let us fix Darboux coordinates $(x_1, \ldots, x_{n-1}, y_1, \ldots, y_{n-1}, z)$ on M such that the contact structure is given by the 1-form

$$\alpha = \sum_{i=1}^{n-1} \frac{x_i\,\mathrm{d}y_i - y_i\,\mathrm{d}x_i}{2} + \mathrm{d}z$$

and compute the explicit formula of the Poisson bracket.

Exercise 7.5.4 Check that, in the above coordinate system, the bracket (7.22) is given by

$$\{\phi, \psi\} = \sum_{i=1}^{n-1} (\phi_{x_i} \psi_{y_i} - \psi_{x_i} \phi_{y_i}) + \phi_z (\mu \psi + \mathcal{E} \psi) - \psi_z (\lambda \phi + \mathcal{E} \phi)$$

where

$$\mathcal{E} = \sum_{i=1}^{n-1} \left(x_i \frac{\partial}{\partial x_i} + y_i \frac{\partial}{\partial y_i} \right)$$

is the Euler field.

The Poisson bracket (7.22) obviously satisfies the Jacobi and Leibnitz identities.

Contact Hamiltonians as tensor densities

The space $\mathcal{F}_{-1/n}(M)$ is a Lie subalgebra of $\mathcal{F}(M)$. Let us show that this algebra is isomorphic to the Lie algebra, $\mathrm{Vect}_c(M)$, of contact vector fields on M.

The isomorphism can be defined as follows. Fix a contact form α, and associate with every contact vector field X a function $F_X = \alpha(X)$. The function F_X depends on the choice of the contact form α but the tensor density

$$\phi_X = \alpha(X)\,\alpha^{-1} \tag{7.23}$$

does not. Note that, to make sense of α^{-1} in the above formula, we understand α as a tensor density of degree $1/n$; cf. Proposition 7.5.2. Thus, ϕ_X is a tensor density of degree $-1/n$. We call this tensor density a *contact Hamiltonian* of X.

Proposition 7.5.5 *The map* $\mathrm{Vect}_c(M) \to \mathcal{F}_{-\frac{1}{n}}(M)$ *sending* X *to* ϕ_X *is an isomorphism of Lie algebras.*

Proof A tensor density of degree $-1/n$ is identified with a homogeneous function on S of degree 1; see Corollary 7.5.3. An arbitrary vector field on M is identified with a fiber-wise linear function on T^*M; this is an isomorphism of Lie algebras. The tensor density (7.23) is just the restriction of this function to S. □

Remark 7.5.6 A contact Hamiltonian is a tensor density of degree $-1/n$ on M, and not a function (cf. [10, 15]). This viewpoint of course resolves the unfortunate difficulty that "the Poisson bracket of two contact Hamiltonians does not satisfy the Leibnitz identity".

Contact manifolds with projective structure

The standard contact structure on $\mathbb{RP}^{2n-1}$ is obtained as the projectivization of the standard symplectic structure on $\mathbb{R}^{2n}$. The "projectivized" symplectic group is

$$\mathrm{PSp}(2n, \mathbb{R}) = \mathrm{Sp}(2n, \mathbb{R})/\mathbb{Z}_2.$$

This is a subgroup of the group of projective transformations $\mathrm{PGL}(2n, \mathbb{R})$ that preserves the contact structure.

Let M be a $(2n-1)$-dimensional manifold equipped with a projective structure. The projective structure is said to be *compatible* with the contact structure if it is given by a projective atlas with the transition functions $\varphi_i \circ \varphi_j^{-1}$ preserving the contact structure on $\mathbb{RP}^{2n-1}$, that is, belonging to the subgroup $\mathrm{PSp}(2n, \mathbb{R})$. In this case, M is a contact manifold with the contact structure induced from $\mathbb{RP}^{2n-1}$.

In the same way, if X is a $2n$-dimensional manifold with an affine structure, one can ask if this affine structure is compatible with the symplectic structure. This means that the transition functions of the affine structure belong to the affine symplectic group $\mathrm{Sp}(2n, \mathbb{R}) \ltimes \mathbb{R}^{2n} \subset \mathrm{Aff}(2n, \mathbb{R})$.

Let us describe a different version of symplectization of a contact manifold with a compatible projective structure. We seek a symplectization which can be naturally equipped with an affine structure.

Example 7.5.7 The model example is the (tautological) line bundle

$$\mathbb{R}^{2n} \setminus \{0\} \to \mathbb{RP}^{2n-1}. \tag{7.24}$$

The group $\mathrm{PSp}(2n, \mathbb{R})$ can be lifted (in two different ways) as a group $\mathrm{Sp}(2n, \mathbb{R})$ of linear symplectic transformations of $\mathbb{R}^{2n}$.

Moreover, one has a stronger statement.

Exercise 7.5.8 Prove that:

(a) the quotient of the group of all homogeneous symplectomorphisms of $\mathbb{R}^{2n} \setminus \{0\}$ by its center $\mathbb{Z}_2$ is isomorphic to $\mathrm{Diff}_c(\mathbb{RP}^{2n-1})$;

(b) the space $\mathcal{F}_\lambda(\mathbb{RP}^{2n-1})$ is isomorphic, as a $\mathrm{Diff}_c(\mathbb{RP}^{2n-1})$-module, to the space of homogeneous functions on $\mathbb{R}^{2n} \setminus \{0\}$ of degree $-2\lambda\, n$.

Let us describe an analogous construction for an arbitrary orientable contact manifold M with a compatible projective structure. Locally M is identified with $\mathbb{RP}^{2n-1}$. One then constructs a line bundle $\widetilde{S} \to M$ such that $\widetilde{S}$ is a symplectic manifold with a compatible affine structure. Locally $\widetilde{S}$ is as (7.24) and the projective transition functions $\varphi_i \circ \varphi_j^{-1}$ can be lifted as elements of $\mathrm{Sp}(2n, \mathbb{R})$.

Let us now relate the two version of symplectization, S and $\widetilde{S}$. The idea comes from the description of a projective structure in terms of tensor densities; see Section 6.4. In the case of manifolds of dimension $2n - 1$, one should consider tensor densities of degree $-1/(2n)$. It is then natural to consider the line bundle $S^{\otimes 2}$ over M.

Proposition 7.5.9 *The bundle $\widetilde{S}$ over an orientable contact manifold M with a compatible projective structure is isomorphic to $S^{\otimes 2}$.*

Proof By construction, the space of homogeneous functions on $\widetilde{S}$ of degree 1 is isomorphic to the space of tensor densities on M of degree $-1/(2n)$. The space of sections of $\widetilde{S}$ is then isomorphic to the space of tensor densities of degree $1/(2n)$. □

Multi-dimensional transvectants and star-products

Let M be again an orientable contact manifold with a compatible projective structure.

Definition 7.5.10 For every m we define a bilinear $\mathrm{PSp}(2n, \mathbb{R})$-invariant differential operator of order $2m$

$$J_m : \mathcal{F}_\lambda(M) \otimes \mathcal{F}_\mu(M) \to \mathcal{F}_{\lambda+\mu+m/n}(M). \tag{7.25}$$

The definition is similar to that of Section 3.1. We identify $\mathcal{F}_\lambda(M)$ with the space of homogeneous functions on $\widetilde{S}$ of degree $-2\lambda\, n$ and consider the iterated Poisson bracket: $B_m := \frac{1}{m!} \mathrm{Tr} \circ P^m$. The operators (7.25) are then given by the restrictions

$$J_m = B_m \left|_{\mathcal{F}_\lambda(M) \otimes \mathcal{F}_\mu(M)}\right. .$$

Consider the space of all smooth functions on $\mathbb{R}^{2n} \setminus \{0\}$. As in the one-dimensional case, formula (3.23) defines an $\mathrm{PSp}(2n, \mathbb{R})$-invariant star-product on this space, called the *Moyal product.* The restriction of this product to the space of homogeneous functions defines a star-product on the Poisson algebra $\mathcal{F}(M)$. We denote by $\mathcal{F}(M)\,[[t]]$ the (associative/Lie) algebra equipped with this star-product.

Extensions of $\mathrm{Vect}_c(M)$

Let us use the following general fact. Given a formal deformation of a Lie algebra, for every subalgebra, one obtains a series of extensions.

We consider the Lie subalgebra of contact vector fields

$$\mathcal{F}_{-1/n}(M) \cong \mathrm{Vect}_c(M)$$

of $\mathcal{F}(M)$, viewed as a Lie algebra. The star-commutator (see formula (3.25)) defines a series of extensions of this Lie algebra.

The first extension is as follows:

$$0 \longrightarrow \mathcal{F}_{1/n}(M) \longrightarrow \mathfrak{g}_1 \longrightarrow \mathrm{Vect}_c(M) \longrightarrow 0; \qquad (7.26)$$

it is given by the 2-cocycle

$$J_3 : \mathcal{F}_{-1/n}(M) \otimes \mathcal{F}_{-1/n}(M) \to \mathcal{F}_{1/n}(M),$$

which is the first non-trivial term of the star-commutator. This is a non-trivial extension of $\mathrm{Vect}_c(M)$ by the module of contact 1-forms.

More generally, one obtains a series of Lie algebras

$$0 \longrightarrow \mathcal{F}_{(2k-1)/n}(M) \longrightarrow \mathfrak{g}_k \longrightarrow \mathfrak{g}_{k-1} \longrightarrow 0, \qquad (7.27)$$

which are consecutive non-trivial extensions.

Example 7.5.11 As a vector space, the Lie algebra $\mathfrak{g}_2$ is as follows

$$\mathfrak{g}_2 = \mathrm{Vect}_c(M) \oplus \mathcal{F}_{1/n}(M) \oplus \mathcal{F}_{3/n}(M).$$

The commutator in $\mathfrak{g}_2$ is given by

$$\left[\begin{pmatrix} X \\ \alpha \\ \phi \end{pmatrix}, \begin{pmatrix} Y \\ \beta \\ \psi \end{pmatrix}\right] = \begin{pmatrix} [X, Y] \\ L_X\beta - L_Y\alpha + J_3(X, Y) \\ L_X\psi - L_Y\phi + \{\alpha, \beta\} + J_3(X, \beta) - J_3(Y, \alpha) + J_5(X, Y) \end{pmatrix}$$

Central extensions: contact Virasoro algebra

The main feature of the Virasoro algebra is that it is defined as a central extension. It turns out that the above defined Lie algebras of tensor densities $\mathcal{F}(M)\,[[t]]$ and $\mathfrak{g}_k$ can also have non-trivial central extensions.

Theorem 7.5.12 *(i) The Lie algebra $\mathcal{F}(M)[[t]]$ has a non-trivial central extension with coefficients in $\mathbb{R}[t]$.*
(ii) If $\dim M = 4k+1$, then the Lie algebra $\mathfrak{g}_k$ has a non-trivial central extension.

Proof Let us construct a non-trivial 2-cocycle on $\mathcal{F}(M)[[t]]$ and on $\mathfrak{g}_k$.

Choose an arbitrary homogeneous function ρ on $\widetilde{S}$, of any degree $\lambda \neq 0$, such that $\rho = 1$ on M. Then $\rho \neq 1$ on $\widetilde{S} \setminus M$. Define first a linear map

$$\gamma : \mathcal{F}(M)[[t]] \to \mathcal{F}(M)[[t]],$$

where $\mathcal{F}(M)[[t]]$ is the algebra of tensor densities with the star-commutator, as follows:

$$\gamma(\phi) = [\phi, \ln \rho] \tag{7.28}$$

where the commutator on the right hand side is the star-commutator (3.25). Note that $\ln \rho$ is a function on $\widetilde{S}$ which is not homogeneous and therefore is not a tensor density on M. However, the commutator (7.28) is an element of $\mathcal{F}(M)[[t]]$. Clearly γ is a 1-cocycle since it is written in the form of "coboundary"; cf. Appendix A.5.

There is an invariant linear functional

$$\int : \mathcal{F}_1(M) \to \mathbb{R}.$$

Define the residue on $\mathcal{F}(M)$ as the integral of the projection $\mathcal{F}(M) \to \mathcal{F}_1(M)$, written as $\phi \mapsto \phi_1$:

$$\operatorname{res}(\phi) = \int \phi_1.$$

For two elements $\phi, \psi \in \mathcal{F}(M)[[t]]$, we set

$$c(\phi, \psi) = \operatorname{res}(\phi\, \gamma(\psi)). \tag{7.29}$$

Exercise 7.5.13 Check that the bilinear form c is skew-symmetric.

Hint. Use the Jacobi identity and the fact that for every $\phi \in \mathcal{F}_\lambda(M)$ and $\psi \in \mathcal{F}_\mu(M)$ with $\lambda + \mu = 1 - 1/n$, one has

$$\int \{\phi, \psi\} = 0.$$

Exercise 7.5.14 Check that c is a 2-cocycle on $\mathcal{F}(M)[[t]]$ with coefficients in the space of polynomials $\mathbb{R}[t]$.

Hint. This is an infinitesimal version of Exercise A.5.6.

We thus defined a central extension

$$0 \longrightarrow \mathbb{R}[t] \longrightarrow \widehat{\mathcal{F}(M)} \longrightarrow \mathcal{F}(M) \longrightarrow 0.$$

We will now restrict the above construction to the Lie algebra $\mathfrak{g}_k$ and put $t = 1$. As a vector space,

$$\mathfrak{g}_k = \bigoplus_{i=0}^{k} \mathcal{F}_{(2i-1)/n},$$

and it can be viewed as a subspace of $\mathcal{F}(M)$.

Exercise 7.5.15 Check that the restriction of the cocycle (7.29) to this subspace is given explicitly by

$$c(\phi_i, \phi_j) = \int \left(\phi_i \, J_{n-2(i+j-1)}(\phi_j, \rho)\right) \tag{7.30}$$

where $\phi_i \in \mathcal{F}_{(2i-1)/n}$.

The cocycle (7.29) is non-trivial since it depends on the $n - 2(i + j - 1)$-jets of ϕ_i and ϕ_j, while any coboundary is a linear function of the commutator $[\phi_i, \phi_j]$ and depends only on the $n - 2(i + j)$-jet. □

In the one-dimensional case, $M = S^1$, the Lie algebra $\mathfrak{g}_1$ is Vect(S^1) itself.

Exercise 7.5.16 Check that, in the one-dimensional case, the defined central extension of the Lie algebra $\mathfrak{g}_1$ is precisely the Virasoro algebra.

In the general case, the central extension of the Lie algebra $\mathfrak{g}_1$ can therefore be considered a multi-dimensional (contact) analog of the Virasoro algebra.

Comment

Unlike the Lie algebra of Hamiltonian vector fields in symplectic geometry, the Lie algebra of contact vector fields $\text{Vect}_c(M)$ is rigid; see [67]. This Lie algebra is traditionally identified with the space $C^\infty(M)$ and then $C^\infty(M)$ is not a Poisson algebra; see [113] for an excellent discussion. It is therefore very important to consider the full space of tensor densities $\mathcal{F}(M)$; this was done in [162].

The first of the series of extensions of $\text{Vect}_c(M)$, namely, the extension (7.26) by the module of contact 1-forms, goes back to Lichnerowicz [136]. The series of extensions (7.27) was defined in [162], the idea of central extensions was also suggested in [162] but with a wrong 2-cocycle. This mistake was corrected and the central extensions were defined in [170]. The star-product on the space of tensor densities on an arbitrary contact manifold was also constructed in [170].

7.6 Lagrange Schwarzian derivative

Relations between projective and symplectic geometries remain somewhat mysterious. However, it is not just by accident that symplectic and contact geometry is extensively used in this book. In this section the symplectic viewpoint occupies a key position. We describe a multi-dimensional symplectic analog of the classic cross-ratio and the Schwarzian derivative.

The main idea is very simple: a point of the projective line $\mathbb{RP}^1$, that is, a one-dimensional subspace of $\mathbb{R}^2$, can be viewed as a Lagrangian subspace. We replace $\mathbb{R}^2$ by the standard symplectic space $(\mathbb{R}^{2n}, \omega)$ and $\mathbb{RP}^1$ by Λ_n, the manifold of Lagrangian subspaces in $(\mathbb{R}^{2n}, \omega)$. This manifold is called the *Lagrange Grassmannian*. The group naturally acting on Λ_n is the group $\mathrm{Sp}(2n, \mathbb{R})$ of linear symplectic transformations; it coincides with $\mathrm{SL}(2, \mathbb{R})$ for $n = 1$.

Arnold–Maslov index

Consider three Lagrangian subspaces (ℓ_1, ℓ_2, ℓ_3) in $(\mathbb{R}^{2n}, \omega)$ such that ℓ_2 and ℓ_3 are transversal. They define a quadratic form on ℓ_1. Namely, every vector $v_1 \in \ell_1$ has a unique decomposition $v_1 = v_2 + v_3$ with $v_2 \in \ell_2$ and $v_3 \in \ell_3$. Define a quadratic form on ℓ_1 by

$$\Phi[\ell_1, \ell_2, \ell_3](v_1) = \omega(v_2, v_3). \qquad (7.31)$$

Clearly, the quadratic form (7.31) is $\mathrm{Sp}(2n, \mathbb{R})$-invariant.

Exercise 7.6.1 Show that the index of the quadratic form $\Phi[\ell_1, \ell_2, \ell_3]$ is the unique linear symplectic invariant of the triple (ℓ_1, ℓ_2, ℓ_3).

The *Arnold–Maslov index* of the triple (ℓ_1, ℓ_2, ℓ_3) is the index of the quadratic form $\Phi[\ell_1, \ell_2, \ell_3]$.

Local coordinates on Λ_n

The Lagrange Grassmannian can be locally identified with the space of symmetric $(n \times n)$-matrices. Let us fix transversal Lagrangian spaces ℓ_1 and ℓ_2 and choose Darboux coordinates (see Appendix A.2) in such a way that ℓ_1 and ℓ_2 become the x-plane and y-plane, respectively. One then associates with every Lagrangian space ℓ, transversal to ℓ_2, a symmetric matrix, representing the quadratic form $\Phi[\ell_1, \ell_2, \ell]$ in x-coordinates on ℓ_1. This defines a local chart on Λ_n analogous to an affine parameterization of $\mathbb{RP}^1$.

Exercise 7.6.2 The action of $\mathrm{Sp}(2n, \mathbb{R})$ is given by

$$X \mapsto (aX + b)(cX + d)^{-1} \tag{7.32}$$

where X is a symmetric $(n \times n)$-matrix and

$$\begin{pmatrix} a & b \\ c & d \end{pmatrix} \in \mathrm{Sp}(2n, \mathbb{R}),$$

that is, $b^* a$ and $c^* d$ are symmetric and $a^* d - b^* c = \mathrm{Id}$; here * denotes transposition of matrices and Id the unit $(n \times n)$-matrix.

Useful formula

Consider a triple of Lagrangian subspaces (ℓ_1, ℓ_2, ℓ_3). Choosing a local chart on Λ_n as above, one represents these subspaces by three symmetric matrices A, B, C. The quadratic form $\Phi[\ell_1, \ell_2, \ell_3]$ then can be expressed in terms of these matrices. For the sake of simplicity, let us assume that ℓ_1 coincides with the x-plane in the Darboux basis; this means that $A \equiv 0$.

Exercise 7.6.3 Check that the quadratic form $\Phi[\ell_1, \ell_2, \ell_3]$ on ℓ_1, in the chosen Darboux basis, is of the form $\langle v, Fv \rangle$ where F is the following $n \times n$-symmetric matrix:

$$F = \left(B^{-1} + C^{-1}\right)^{-1} \tag{7.33}$$

One can generalize this formula for an arbitrary A.

Symplectic cross-ratio

Consider now a quadruple $(\ell, \ell_1, \ell_2, \ell_3)$ of Lagrangian subspaces in $(\mathbb{R}^{2n}\omega)$. The pair of quadratic forms on ℓ

$$(\Phi[\ell, \ell_1, \ell_3],\ \Phi[\ell, \ell_2, \ell_3]) \tag{7.34}$$

is the unique linear symplectic invariant of the quadruple. We denote this pair of quadratic forms by $[\ell, \ell_1, \ell_2, \ell_3]$ and call it the *symplectic cross-ratio*.

Exercise 7.6.4 Check that, in the one-dimensional case, the cross-ratio (1.5) of four lines in $\mathbb{R}^2$ is the quotient of the above defined forms:

$$[\ell, \ell_1, \ell_2, \ell_3] = \frac{\Phi[\ell, \ell_2, \ell_3]}{\Phi[\ell, \ell_1, \ell_3]}. \tag{7.35}$$

Remark 7.6.5 In the one-dimensional case, expression (7.35) is always well-defined for four distinct lines. In the multi-dimensional case, the form $\Phi[\ell, \ell_1, \ell_2]$ is degenerate if ℓ is not transversal to ℓ_1 or ℓ_2. This is why it is natural to deal with the couple (7.34), rather than with, say, their ratio.

Assume that $\Phi[\ell, \ell_1, \ell_3]$ is non-degenerate and choose a basis of ℓ such that $\Phi[\ell, \ell_1, \ell_3] = \mathrm{diag}(1, \ldots, 1, -1, \ldots, -1)$. Then one can think of the quadratic form $\Phi[\ell, \ell_2, \ell_3]$ as a symmetric $(n \times n)$-matrix, defined up to orthogonal transformations: $\Phi \sim O\,\Phi\,O^{-1}$.

Non-degenerate curves in Λ_n

Consider a smooth map

$$f : \mathbb{RP}^1 \to \Lambda_n. \tag{7.36}$$

We will construct a differential invariant of maps (7.36) with respect to the $\mathrm{Sp}(2n, \mathbb{R})$-action on the Lagrange Grassmannian. For $n = 1$, one has $\Lambda_n = \mathbb{RP}^1$, and this invariant coincides with the classic Schwarzian derivative.

We will assume that the matrix $f'(x)$ is non-degenerate for every x. This condition has a geometrical meaning. With every point ℓ of Λ_n we associate a codimension one variety (with singularity in ℓ), called the *train*. The train consists of Lagrangian spaces which are not transversal to ℓ. The corresponding curve in Λ_n is everywhere transversal to the train; see figure 7.1. We call a map (7.36) satisfying this condition *non-degenerate*. The image of $\mathbb{RP}^1$ under a non-degenerate map is called a non-degenerate curve in Λ_n. In the one-dimensional case, the non-degeneracy condition means that f is a (local) diffeomorphism.

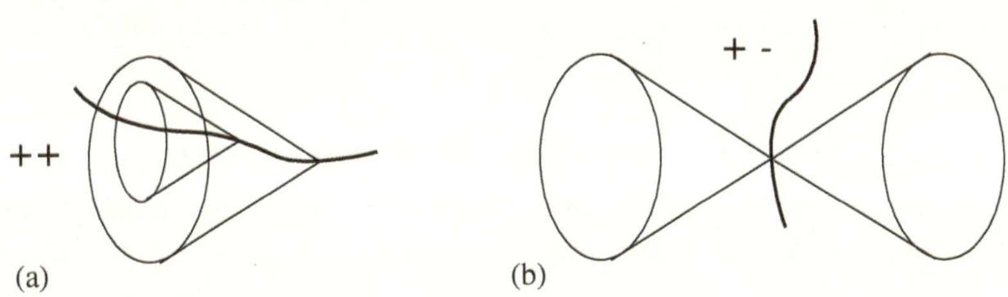

Fig. 7.1. Non-degenerate curves in Λ_n

Infinitesimal symplectic cross-ratio

The main idea is similar to the one-dimensional case; see Section 1.3. Let x be a point in $\mathbb{RP}^1$ and v a tangent vector to $\mathbb{RP}^1$ at x. We extend v to a vector field in a vicinity of x and denote by ϕ_t the corresponding local one-parameter group of diffeomorphisms. We then consider four points:

$$x, \quad x_1 = \phi_\varepsilon(x), \quad x_2 = \phi_{2\varepsilon}(x), \quad x_3 = \phi_{3\varepsilon}(x)$$

(ε is small) and compute the symplectic cross-ratio of their images under f. More precisely, we compute the pair of quadratic forms:

$$\Phi_1 = \Phi[f(x), f(x_1), f(x_3)], \qquad \Phi_2 = \Phi[f(x), f(x_2), f(x_3)].$$

Choosing, as above, an atlas on Λ_n, one represents the family of Lagrangian subspaces $f(x)$ as a family of symmetric $(n \times n)$-matrices. Each of the above quadratic forms will also be represented by a family of symmetric matrices.

Exercise 7.6.6 Let $f(x)$ be a non-degenerate curve in Λ_n. Show that, for every local chart on Λ_n, the symplectic cross-ratio is given by two symmetric matrices

$$\begin{aligned} \Phi_1 &= \tfrac{3}{2}\,\varepsilon\, f' - \tfrac{3}{4}\varepsilon^3 \left(f''' - \tfrac{3}{2}\, f''\left(f'\right)^{-1} f''\right) + O(\varepsilon^4), \\ \Phi_2 &= 6\,\varepsilon f' + 6\,\varepsilon^3 \left(f''' - \tfrac{3}{2}\, f''\left(f'\right)^{-1} f''\right) + O(\varepsilon^4). \end{aligned} \tag{7.37}$$

Hint. Use the Taylor expansion for $f(x_1)$, $f(x_2)$ and $f(x_3)$; for instance,

$$f(x_1) = f(x) + \varepsilon\, f'(x) + \frac{\varepsilon^2}{2}\, f''(x) + \frac{\varepsilon^3}{6}\, f'''(x) + O(\varepsilon^4).$$

One can then assume, without loss of generality, that $f(x) = 0$ and apply formula (7.33).

Remark 7.6.7 The reader will already have noticed similarity between expression (7.37) and the classic Schwarzian derivative (1.8). In the one-dimensional case, it suffices to take the quotient Φ_2/Φ_1 to obtain formula (1.6).

Introducing Lagrange Schwarzian derivative

Assume that the non-degenerate map f is *positive*, that is, the matrix $f'(x)$ is positive definite. An example is given by Figure 7.1, case (a). The families of matrices $\Phi_1(x)$ and $\Phi_2(x)$ are then positive definite for ε small enough. There is a family of matrices $C(x)$, not necessarily symmetric, satisfying the condition $\Phi_1(x) = C^*(x)C(x)$. Any such family $C(x)$ will be called a *square root* of $\Phi_1(x)$ and denoted by $\sqrt{\Phi_1(x)}$. The square root is defined modulo orthogonal matrices:

$$\sqrt{\Phi_1(x)} \sim O(x)\sqrt{\Phi_1(x)}.$$

Choosing a basis in the Lagrangian space $f(x)$ which is orthonormal for the form Φ_1, one obtains for the symplectic cross-ratio

$$\Phi_1 = \mathrm{Id}, \qquad \Phi_2 = 4\,\mathrm{Id} - \tfrac{1}{2}\,\varepsilon^2\,LS(f) + O(\varepsilon^4)$$

where

$$LS(f) = \sqrt{(f')^{-1}}\left(f''' - \tfrac{3}{2}\,f''\,(f')^{-1}\,f''\right)\sqrt{(f')^{-1}}^{\,*}. \tag{7.38}$$

This expression will be called the *Lagrange Schwarzian derivative*.

Note that expression (7.38) is defined up to conjugation by families of orthogonal matrices:

$$LS(f(x)) \sim O(x)\,LS(f(x))\,O(x)^{-1}.$$

We will define, in the next subsection, the notion of canonical square root and the Lagrange Schwarzian will be defined up to conjugation by orthogonal matrices which do not depend on the parameter.

Proposition 7.6.8 *One has* $LS(f(x)) = LS(g(x))$ *if and only if* $g(x) = (af(x)+b)\,(cf(x)+d)^{-1}$ *where* $\begin{pmatrix} a & b \\ c & d \end{pmatrix}$ *is a symplectic matrix.*

Proof This follows from the symplectic invariance of the quadratic forms Φ_1 and Φ_2. □

Square root of symmetric matrices and loop group action

With an arbitrary family $G(x)$ of non-degenerate $(n \times n)$-matrices we will associate another family:

$$c(G) = G'(x)\,G(x)^{-1}. \tag{7.39}$$

The algebraic meaning of the map c is as follows. Let $\mathfrak{G} = C^\infty(S^1, \mathrm{GL}(n, \mathbb{R}))$ be the group of smooth functions with pointwise multiplication, usually called the *loop group*. Let $\mathfrak{g}$ be the corresponding Lie algebra, that is, $\mathfrak{g} = C^\infty(S^1, \mathrm{gl}(n, \mathbb{R}))$. The (co)adjoint action of $\mathfrak{G}$ on $\mathfrak{g}$ is given by

$$\mathrm{ad}_{G(x)} M(x) = G(x)\, M(x)\, G(x)^{-1}.$$

The map (7.39) is a 1-cocycle on $\mathfrak{G}$ with coefficients in $\mathfrak{g}$ so that

$$\widetilde{\mathrm{ad}}_G = \mathrm{ad}_G + c(G)$$

is an affine action; see Appendix A.4.

Proposition 7.6.9 *Let $X(x)$ be a family of symmetric positively definite $(n \times n)$-matrices.*
(i) There is a family of matrices $B(x)$ such that:

$$(1) \quad B^* B = X, \qquad (2) \quad c(B) \quad \textit{is symmetric for all } x.$$

(ii) The family $B(x)$ is uniquely defined up to multiplication by a constant orthogonal matrix: $B(x) \sim O\, B(x)$.

Proof Let $B(x)$ be an arbitrary family of matrices satisfying condition (1). Consider the equation $O'(x) = O(x)\,(c(B)^* - c(B))$. This equation has a unique solution $O(x) \in \mathrm{SO}(n)$, defined up to multiplication: $O(x) \sim \widetilde{O}\, O(x)$. The corresponding family $O(x)\, B(x)$ satisfies condition (2). □

We call the family $B(x)$ defined in Proposition 7.6.9 the *canonical square root* of a family of symmetric matrices.

We will understand the square roots in (7.38) as the canonical ones. The Lagrange Schwarzian derivative is then uniquely defined up to conjugation by constant orthogonal matrices: $LS(f(x)) \sim O\, LS(f(x))\, O^{-1}$.

Lagrange Schwarzian derivative and Newton systems

The classic Schwarzian derivative is closely related to the Sturm–Liouville operators; see Section 1.3. In the multi-dimensional case, analogs of the Sturm–Liouville operators are second-order matrix differential operators:

$$L = \frac{\mathrm{d}^2}{\mathrm{d}x^2} + A(x) \tag{7.40}$$

where $A(x)$ is a family of symmetric $(n \times n)$-matrices. We call operators (7.40) *Newton operators*.

Consider the Newton system of n linear differential equations

$$y''(x) + A(x)\, y(x) = 0 \tag{7.41}$$

where $y \in \mathbb{R}^n$. The $2n$-dimensional space of its solutions has a natural symplectic structure given by an analog of the Wronski determinant:

$$W(y, \tilde{y}) = \sum_{i=1}^{n} \left(y_i(x)\, \tilde{y}_i'(x) - y_i'(x)\, \tilde{y}_i(x) \right).$$

Exercise 7.6.10 Check that this expression does not depend on x and defines a non-degenerate skew-symmetric form on the space of solutions.

The space of solutions is therefore isomorphic to the standard symplectic space $(\mathbb{R}^{2n}, \omega)$.

Every Newton operator defines a positive non-degenerate curve in Λ_n. Indeed, let $\ell(x)$ be the n-dimensional subspace of solutions vanishing at point x; this space is clearly Lagrangian.

Proposition 7.6.11 *The curve $\ell(x)$ is positive.*

Proof Let $y_1(x), \ldots, y_n(x) \in \mathbb{R}^n$ be linearly independent solutions of (7.41). Denote by $Y(x)$ the $(n \times n)$-matrix whose i-th column is the vector y_i. For every two matrices of solutions $Y(x)$ and $Z(x)$ we define their Wronskian

$$W(Y, Z) = Y(x)^*\, Z'(x) - Y'(x)^*\, Z(x),$$

which is a constant matrix. It is clear that $W_{ij} = W(y_i, z_j)$. The space of solutions is a $2n$-dimensional symplectic space. Consider a Darboux basis $\{y_1, \ldots, y_n, z_1, \ldots, z_n\}$ in this space and the corresponding matrices of solutions $Y(x)$ and $Z(x)$. Then one has

$$W(Y, Z) = \mathrm{Id}, \qquad W(Y, Y) = 0, \qquad W(Z, Z) = 0. \tag{7.42}$$

Denote by α and β the Lagrange subspaces in the space of solutions corresponding to the matrices $Y(x)$ and $Z(x)$.

Consider the Lagrange subspace $\ell(s)$ in the space of solutions. Let $L_s(x)$ be a fundamental matrix of solutions corresponding to $\ell(s)$ (i.e., we choose a basis in $\ell(s)$). Then, up to a conjugation,

$$L_s(x) = Y(x)\, Y(s)^{-1} - Z(x)\, Z(s)^{-1}.$$

The quadratic form $\Phi[\alpha, \beta, \ell(s)]$ is thus given, in the chosen basis, by the family of symmetric $(n \times n)$-matrices $f(s) = Y(s)^{-1}\, Z(s)$, that is, $Z(s) = Y(s)\, f(s)$.

Substituting this expression in (7.42), one obtains $Y(s)^* \, Y(s) = f'(s)^{-1}$. It follows that $f'(s)$ is positively definite for every s. □

Note, furthermore, that the matrix $Y(x)$ is the canonical square root:

$$Y(x) = \sqrt{f'(x)^{-1}}.$$

Indeed, we already proved that it satisfies condition (1) in Proposition 7.6.9.

Exercise 7.6.12 Check that the matrix $Y'(x) \, Y(x)^{-1}$ is symmetric if and only if $W(Y, Y) = 0$.

That is, condition (2) of the canonical square root is equivalent to the fact that α is a Lagrangian plane.

Let us now consider the inverse problem. Given a positive non-degenerate curve in Λ_n, is it possible to find an operator (7.40) corresponding to this curve? It turns out the the answer in provided by the Lagrange Schwarzian derivative. The following statement is an analog of Exercise 1.3.4.

Theorem 7.6.13 *Every positive non-degenerate curve $f(x)$ in Λ_n corresponds to a Newton operator with*

$$A = \frac{1}{2} \, LS(f(x)).$$

Proof Assume that a curve $f(x) \subset \Lambda_n$ corresponds to some Newton system (7.41) in the same sense as above. We would like to calculate explicitly the potential $A(x)$. Fix a Darboux basis in the space of solutions, as in the proof of Proposition 7.6.11. This defines a local chart on Λ_n, and one associates a family of symmetric $(n \times n)$-matrices to $f(x)$. We already proved that $Y(x) = \sqrt{f'(x)^{-1}}$ is a fundamental matrix of solutions of (7.41). But the potential can then be easily computed as $A(x) = Y''(x) \, Y(x)^{-1}$.

Exercise 7.6.14 Check that $LS(f(x)) = 2 \, Y''(x) \, Y(x)^{-1}$.

The computation does not depend on the choice of the Darboux basis since, for a fixed $(n \times n)$-matrix C, one has $LS(C^* \, f(x) \, C) = LS(f(x))$; and this computation can be made for any positive non-degenerate curve $f(x)$. This completes the proof. □

Comment

The idea to associate an evolution of a Lagrangian subspace to a Newton system is due to V. Arnold [7]. In particular, he proved that the corresponding curve in Λ_n is positive. The exposition in this section is based on [165].

As we mentioned in the Preface, the classic Schwarzian derivative was, most likely, invented by Lagrange. It seems very appropriate that Lagrange's name is again connected to (a version of) the Schwarzian derivative at the very end of this book!

Appendices

A.1 Five proofs of the Sturm theorem

The classic Sturm theorem states that the number of zeroes of a periodic function is not less than that of its first non-trivial harmonic. This theorem is intimately related to inflections of projective curves, theorems on vertices of plane curves and other results discussed in Chapter 3.

Believing in the maxim that it is better to prove the same result in many different ways than to prove different results in the same way, we devote this section to a number of different proofs of the Sturm theorem.

Formulation of the Sturm theorem

The Sturm theorem provides a lower bound for the number of zeroes of a smooth 2π-periodic function $g(x)$ whose Fourier expansion

$$g(x) = \sum_{k \geq n} (a_k \cos kx + b_k \sin kx) \tag{A.1}$$

starts with harmonics of order n.

Theorem A.1.1 *The function $g(x)$ has at least $2n$ distinct zeroes on the circle $[0, 2\pi)$.*

In the first two proofs, we assume that $g(x)$ is a trigonometric polynomial (of degree N); the last three proofs apply to all smooth functions.

One can replace the trigonometric polynomials of degree $\leq n - 1$ by an arbitrary Chebyshev system.

Theorem A.1.2 *Let A be a disconjugate differential operator on S^1 of order $2n - 1$, and let $g(x)$ be a function, orthogonal to all solutions of the differential equation $Af = 0$. Then $g(x)$ has at least $2n$ distinct zeroes.*

The last of our proofs covers this more general result. One can also consider disconjugate differential operators of order $2n$. In this case, the function g has to be anti-periodic (as well as the solutions of $Af = 0$), that is, $g(x + 2\pi) = -g(x)$.

Proof by Barner's theorem

A consequence of the Barner theorem, Theorem 4.4.9, implies that if A is a disconjugate differential operator of order $2n-1$ on the circle and f is a smooth periodic function then the function Af has at least $2n$ distinct zeroes.

Let A be the differential operator A_{n-1} in (4.13). This operator "kills" the harmonics of orders $\le n-1$. Applied to trigonometric polynomials of degree N, the image of A_{n-1} consists precisely of such polynomials that start with harmonics of order n. In particular, $g \in \operatorname{Im} A_{n-1}$, and therefore g has at least $2n$ distict zeroes.

Proof by the argument principle

Consider the complex polynomial

$$G(z) = \sum_{k=n}^{N} (a_k - \mathrm{i}b_k)\, z^k .$$

Then $g(x) = \operatorname{Re} G(\mathrm{e}^{\mathrm{i}x})$. Let γ be the image of the unit circle under G.

Assume first that γ does not pass though the origin. According to the argument principle, the rotation number of γ with respect to the origin equals the number of zeroes of G inside the unit disc (multiplicities counted). Since G has a zero of order n at the origin, this rotation number is $\ge n$. It follows that γ intersects the vertical axis at least $2n$ times, that is, $g(x)$ has at least $2n$ zeroes.

If γ passes though the origin, then we choose sufficiently small $\varepsilon > 0$ and consider the circle of radius $1-\varepsilon$ that does not contain zeroes of G. The image of this circle, γ_ε, still has the rotation number $\ge n$. Thus γ_ε intersects the vertical axis at least $2n$ times and the number of intersections of γ cannot be smaller.

Proof by Rolle theorem

Denote by $Z(f)$ the number of sign changes of $f \in C^\infty(S^1)$. The Rolle theorem asserts that $Z(f') \ge Z(f)$. Introduce the operator D^{-1} on the subspace of functions with zero average:

$$(\mathrm{D}^{-1} f)(x) = \int_0^x f(t)\,\mathrm{d}t.$$

The Rolle theorem then reads: $Z(f) \ge Z(\mathrm{D}^{-1} f)$.

Consider the sequence of functions

$$g_m = (-1)^m \left(n\,\mathrm{D}^{-1}\right)^{2m} g$$

where g is as in (A.1), explicitly,

$$g_m(x) = (a_n \cos nx + b_n \sin nx) + \sum_{k>n} \left(\frac{n}{k}\right)^{2m} (a_k \cos kx + b_k \sin kx)\,. \tag{A.2}$$

By the Rolle theorem, for every m, one has: $Z(g) \ge Z(g_m)$.

Since the Fourier series (A.1) converges, $\sum_k (a_k^2 + b_k^2) < C$ for some constant C. This implies that the second summand in (A.2) is arbitrary small for sufficiently large m. It follows that g_m has at least $2n$ sign changes for large m, and we are done.

Proof by heat equation

Consider the function $g(x)$ as the initial condition for the heat equation

$$\frac{\partial G(x,t)}{\partial t} = \frac{\partial^2 G(x,t)}{\partial x^2}, \qquad G(x,0) = g(x).$$

The number of sign changes of the function $G(x,t)$, considered as a function of x, does not increase with t. This follows from the maximum principle in partial differential equations, but it is also intuitively clear: an iceberg can melt down in a warm sea but cannot appear out of nowhere. On the other hand, one can solve the heat equation explicitly:

$$G(x,t) = \sum_{k \geq n} \mathrm{e}^{-k^2 t} \left(a_k \cos kx + b_k \sin kx\right).$$

The rest of the argument repeats the preceding proof by the Rolle theorem: the higher harmonics tend to zero faster than the n-th ones. Thus, $G(x,t)$ has at least $2n$ zeroes for t large enough.

Proof by orthogonality to trigonometric polynomials

Let us argue by contradiction. Assume that g has less than $2n$ sign changes on the circle. The number of sign changes being even, g has at most $2(n-1)$ of them. One can find a trigonometric polynomial f of degree $\leq n-1$ that changes signs precisely in the same points as g. Then the function fg has a constant sign on the circle and $\int fg\,\mathrm{d}x \neq 0$. But, on the other hand, f is L_2-orthogonal to g; this is a contradiction.

Exercise A.1.3 Let $0 \leq \alpha_1 < \beta_1 < \alpha_2 < \ldots < \beta_{n-1} < 2\pi$ be the points of sign change of the function $g(x)$. Prove that one can set, in the above argument:

$$f(x) = \sin\frac{x-\alpha_1}{2}\sin\frac{x-\beta_1}{2}\cdots\sin\frac{x-\alpha_{n-1}}{2}\sin\frac{x-\beta_{n-1}}{2}.$$

Comment

The Sturm theorem appeared in [195] for trigonometric polynomials; the general case is due to Hurwitz. This theorem has been a source of inspiration for many generations of mathematicians. Over the years, the result has been rediscovered many times; see [12] for the history.

The proofs by the argument principle and by orthogonality can be found in [174], Problems III, 184 and II, 141. The proof by the Rolle theorem appears to be the most recent one [142, 105]; the argument resembles that in [175]. The proof by the heat equation resembles that of the four-vertex theorem by the curve shortening argument in Section 4.6.

Of recent results, we would like to mention [60], in which the Sturm theorem is extended from Fourier series to Fourier integrals, and [218] where analogs of extactic points for trigonometric polynomials are studied.

A.2 The language of symplectic and contact geometry

This section provides a very brief tour of symplectic and contact geometry. Our goal here is to give the reader a general impression of the subject and collect various notions and formulæ which are used many times throughout the book.

Symplectic structure

A *symplectic manifold* (M, ω) is a smooth manifold M with a closed non-degenerate differential 2-form ω, called a *symplectic structure.*

Since the 2-form is non-degenerate, the dimension of a symplectic manifold is even. A symplectic manifold has a canonical volume form ω^n, where $2n = \dim M$.

Example A.2.1 A $2n$-dimensional vector space with linear coordinates $(x_1, \dots, x_n, y_1, \dots, y_n)$ has a linear symplectic structure

$$\omega = \mathrm{d}x_1 \wedge \mathrm{d}y_1 + \cdots + \mathrm{d}x_n \wedge \mathrm{d}y_n. \tag{A.3}$$

Symplectic manifolds of dimension n are all locally equivalent (Darboux's theorem): in a neighborhood of any point there exist local coordinates on M in which the symplectic form is given by (A.3). These coordinates are called *Darboux coordinates.*

The following example is of importance.

Example A.2.2 The *cotangent bundle* T^*N of a smooth manifold N has a symplectic structure. Let λ be the tautological differential 1-form on T^*N, called the *Liouville form*, whose value on a vector v, tangent to T^*N at a point (x, ξ) with $x \in N$ and $\xi \in T_x^*N$, is equal to the value of the covector ξ on the projection of v to the tangent space T_xN. The natural symplectic structure on T^*N is the 2-form $\omega = \mathrm{d}\lambda$.

Choose local coordinates $(x_1, \dots, x_n)$ on N and the dual coordinates $(\xi_1, \dots, \xi_n)$ on T_x^*N. In these coordinates

$$\lambda = \xi_1\, \mathrm{d}x_1 + \cdots + \xi_n\, \mathrm{d}x_n \tag{A.4}$$

and ω is as in (A.3).

An n-dimensional submanifold L of a symplectic manifold (M^{2n}, ω) is called *Lagrangian* if the restriction of ω to L vanishes. Since ω is a non-degenerate 2-form, n is the greatest such dimension.

A diffeomorphism of symplectic manifolds that carries one symplectic structure to another is called a *symplectomorphism.* Symplectomorphisms take Lagrangian manifolds to Lagrangian manifolds.

Exercise A.2.3 If $f : (M_1, \omega_1) \to (M_2, \omega_2)$ is a symplectomorphism then its graph is a Lagrangian submanifold of the product manifold $M_1 \times M_2$ with the symplectic structure $\omega_1 \ominus \omega_2$.

Poisson structures and symplectic leaves

A Poisson structure on a smooth manifold M is a Lie algebra structure $\{.,.\}$ on $C^\infty(M)$ satisfying the Leibnitz identity:

$$\{F, GH\} = \{F, G\}H + G\{F, H\}. \tag{A.5}$$

It follows that the operator of Poisson bracket with a function $\{F, .\}$ is a derivation of the algebra $C^\infty(M)$, that is, a vector field on M. This vector field is called Hamiltonian and denoted by X_F. The correspondence $F \mapsto X_F$ is given by the *Hamiltonian operator* $\mathcal{H} : T^*M \to TM$ via the formula

$$X_F = \mathcal{H}(\mathrm{d}F).$$

This is a homomorphism of Lie algebras: $[X_F, X_G] = X_{\{F,G\}}$.

The basic example of a Poisson manifold is a symplectic manifold, the Poisson bracket being given by $\{F, G\} = \omega(X_F, X_G)$. Moreover, a Poisson structure defines a foliation on M with symplectic leaves. The tangent space to a leaf is spanned by Hamiltonian vector fields.

Another way to define a Poisson structure is to consider a bivector field Λ on M and to define the Poisson bracket by

$$\{F, G\} = \Lambda(\mathrm{d}F, \mathrm{d}G).$$

The Leibnitz identity (A.5) is satisfied automatically, but the Jacobi identity imposes a serious restriction on Λ that we do not specify here.

Given a (finite-dimensional) Lie algebra $\mathfrak{g}$, the dual space $\mathfrak{g}^*$ is endowed with a natural Poisson structure, called the Lie–Poisson bracket, given by the formula

$$\{F, G\}(\varphi) = \langle \varphi, [\mathrm{d}F_\varphi, \mathrm{d}G_\varphi] \rangle \tag{A.6}$$

where $\varphi \in \mathfrak{g}^*$ and the differentials $\mathrm{d}F_\varphi$ and $\mathrm{d}G_\varphi$ are understood as elements of $(\mathfrak{g}^*)^* \cong \mathfrak{g}$. The symplectic leaves of the Lie–Poisson bracket are precisely the coadjoint orbits of $\mathfrak{g}$, endowed with the famous Kirillov symplectic form. Strictly speaking, the above discussion applies to finite-dimensional Lie algebras; however, formula (A.6) often works in the infinite-dimensional case as well.

Symplectic structure on the space of geodesics

Let M^n be a Riemannian manifold. Consider a Hamiltonian function $H : T^*M \to \mathbb{R}$ given by the formula $H(q, p) = |p|^2/2$. The Hamiltonian flow X_H is called the *geodesic flow* on the cotangent bundle. The relation with geodesic lines is as follows: the projection of a trajectory of the geodesic flow in T^*M to M is a geodesic line therein.

Let $N \subset M$ be a smooth hypersurface and q a point of N. Let $p \in T_q^*M$ be a conormal, a covector vanishing on T_qN. Then the projection of the vector $X_H(q, p)$ to T_qM is orthogonal to the hypersurface N.

Let $S \subset T^*M$ be the unit covector hypersurface given by $H = 1/2$. The trajectories of the geodesic flow are tangent to S. Assume that the space of these trajectories is a smooth manifold $\mathcal{L}^{2n-2}$ (locally, this is always the case). Points of $\mathcal{L}$ are non-parameterized oriented geodesics on M. The following *symplectic reduction* construction provides the space of geodesics with a symplectic structure.

Let ω be the canonical symplectic form on T^*M. The restriction of ω on S has a one-dimensional kernel, called the *characteristic direction*, generated by the vector field X_H. Therefore the quotient space $\mathcal{L}$ has a symplectic structure $\bar{\omega}$.

Exercise A.2.4 Consider the case of $M = \mathbb{R}^n$. Prove that the space of oriented lines is symplectomorphic to T^*S^{n-1}. In particular, if $n = 2$ then $\bar{\omega} = \mathrm{d}p \wedge \mathrm{d}\alpha$ where α is the direction of the line and p is the signed distance from the origin to the line.

Hint. An oriented line ℓ is characterized by its unit vector $q \in S^{n-1}$ and the perpendicular vector p, dropped to ℓ from the origin.

Assume that the space of oriented geodesics $\mathcal{L}^{2n-2}$ is a smooth manifold, considered with its symplectic structure $\bar{\omega}$. Let $N \subset M$ be (a germ of) a smooth cooriented hypersurface and $L^{n-1} \subset \mathcal{L}$ the set of geodesics orthogonal to N.

Lemma A.2.5 *L is a Lagrangian submanifold in $\mathcal{L}$. Conversely, a Lagrangian submanifold in $\mathcal{L}$ locally consists of the geodesics orthogonal to a hypersurface in M.*

Proof Identify tangent vectors and covectors on M by the Riemannian metric. Let $p(x)$, $x \in N$, be the field of unit normal vectors along N. Then the 1-form $\langle p, \mathrm{d}x \rangle$ vanishes on N. To prove that L is Lagrangian we need to show that so is the preimage of L in T^*M. Let $q(x, t)$ be the point on the orthogonal geodesic through $x \in N$ at distance t from x; if $M = \mathbb{R}^n$ then $q(x, t) = x + tp(x)$. One parallel translates p to $q(x, t)$. Then

$$\langle p, \mathrm{d}q \rangle = \langle p, \mathrm{d}x \rangle + \mathrm{d}t, \tag{A.7}$$

and therefore $\mathrm{d}p \wedge \mathrm{d}q = 0$. Hence L is Lagrangian.

Conversely, let $L^{n-1} \subset \mathcal{L}$ be a Lagrangian submanifold. Let $N \subset M$ be a local hypersurface, transverse to the geodesics from L, and $p(x)$, $x \in N$, be the unit vector field along N, tangent to these geodesics. We want to show that there exists a function $t(x)$ such that the locus of points $q(x, t)$ is a hypersurface, orthogonal to the lines from L. This is equivalent to the vanishing of the 1-form $\langle p, \mathrm{d}q \rangle$. Since L is Lagrangian, $\mathrm{d}p \wedge \mathrm{d}q = 0$. By (A.7), $\mathrm{d}p \wedge \mathrm{d}x = 0$, and therefore, locally, $\langle p, \mathrm{d}x \rangle = \mathrm{d}f$ for some function on N. Set $t(x) = -f(x)$ and, by (A.7), one has $\langle p, \mathrm{d}q \rangle = 0$. □

Complete integrability

One can discuss complete integrability of a continuous or discrete time dynamical system; to fix ideas, we choose the latter. Let T be a symplectomorphism of a symplectic manifold (M^{2n}, ω). The map T is called *completely integrable* if there exist T-invariant smooth functions $f_1, \dots, f_n$ (integrals) whose pairwise Poisson brackets vanish and which are functionally independent almost everywhere on M; this means that their differentials are linearly independent in an open dense set.

Non-degenerate level sets of the functions $f_1, \dots, f_n$ are Lagrangian manifolds. These manifolds are the leaves of a Lagrangian foliation, leaf-wise preserved by T. The existence of such a foliation can be taken as the definition of complete integrability. By Example 6.1.8, the leaves of a Lagrangian foliation carry a canonical affine structure. The map T, restricted to a leaf, is an affine map. If a leaf is compact then it is a torus, and the restriction of T to this torus is a translation: $x \mapsto x + a$. This statement is (part of) a discrete version of the Arnold–Liouville theorem; see [10, 226].

Contact structures and space of contact elements

Contact geometry is an odd-dimensional counterpart of symplectic geometry. Let M be a $(2n-1)$-dimensional manifold. A *contact structure* on M is a codimension 1 distribution ξ which is completely non-integrable. The distribution ξ can be locally defined as the kernel of a differential 1-form η. Complete non-integrability means that

$$\eta \wedge \underbrace{d\eta \wedge \cdots \wedge d\eta}_{(n-1)\text{ times}} \neq 0 \tag{A.8}$$

everywhere on M.

Example A.2.6 The space $\mathbb{RP}^{2n-1}$ has a canonical contact structure defined as the projectivization of the linear symplectic structure (A.3) in $\mathbb{R}^{2n}$. A point $x \in \mathbb{RP}^{2n-1}$ is a one-dimensional subspace X in $\mathbb{R}^{2n}$. Its orthogonal complement with respect to the symplectic structure is a $(2n-1)$-dimensional subspace containing X. Define the contact hyperplane $\xi(x)$ at x as the projectivization of this hyperplane.

Choose affine coordinates on $\mathbb{RP}^{2n-1}$:

$$u_i = \frac{x_i}{y_n}, \quad v_i = \frac{y_i}{y_n}, \quad w = \frac{x_n}{y_n}, \qquad i = 1, \dots, n-1.$$

Exercise A.2.7 The canonical contact structure in $\mathbb{RP}^{2n-1}$ is the kernel of the 1-form

$$\eta = \sum_{i=1}^{n-1} (u_i dv_i - v_i du_i) - dw. \tag{A.9}$$

Similarly to symplectic manifolds, all contact manifolds of the same dimension are locally diffeomorphic. Locally, any contact form can be written as (A.9).

Example A.2.8 Let N^n be a smooth manifold. A *contact element* in N is a pair (x, H) where x is a point of N and H a hyperplane in $T_x N$. The space of contact elements is a $(2n-1)$-dimensional manifold which can be identified with the projectivization of the cotangent bundle T^*N. This manifold carries a codimension 1 distribution ξ, defined by the following "skating" condition: the velocity vector of the foot point x is tangent to H; see Figure A.1 for $n = 2$.

Exercise A.2.9 Prove that $\mathbb{RP}^3$ is contactomorphic to the space of cooriented contact elements of S^2.

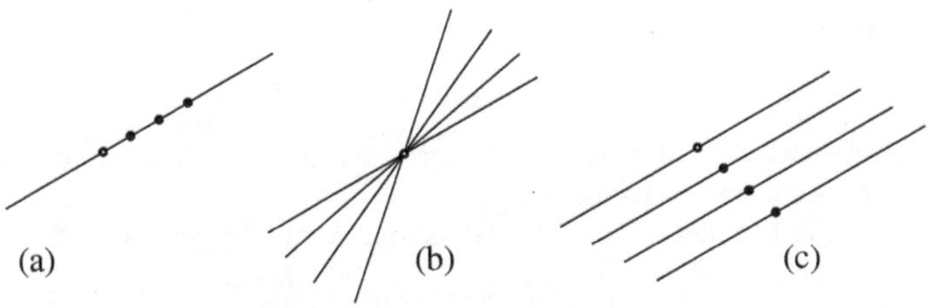

Fig. A.1. Motions (a) and (b) are tangent to the contact plane, motion (c) is transverse to it.

An $(n-1)$-dimensional submanifold L of a contact manifold (M^{2n-1}, ξ) is called *Legendrian* if L is everywhere tangent to ξ. Since ξ is a completely non-integrable, $n-1$ is the greatest such dimension.

If L is a Legendrian submanifold in the space of contact elements in N, its projection to N is called the *front* of L. The front may have singularities. Conversely, any hypersurface in N lifts to a Legendrian submanifold by assigning the tangent hyperplane to its every point.

Example A.2.10 The notion of projective duality of curves in $\mathbb{RP}^2$ was discussed in the very beginning of this book. Let M be the space of contact elements in $\mathbb{RP}^2$, that is, of pairs (x, ℓ) where x is a point and ℓ is a line in $\mathbb{RP}^2$ such that $x \in \ell$. There are two natural projections $p_1 : M \to \mathbb{RP}^2$ and $p_2 : M \to \mathbb{RP}^{2*}$ given by the formulæ $p_1(x, \ell) = x$ and $p_2(x, \ell) = \ell$. Let γ be a curve in $\mathbb{RP}^2$, possibly with cusps, and Γ its Legendrian lift. Then the dual curve is the second projection of Γ, namely $\gamma^* = p_2(\Gamma)$.

Comment

Symplectic and contact geometry is a vast research area, closely related to various branches of contemporary mathematics. The reader unfamiliar with the subject is invited to consult [15, 98, 148]. Symplectic geometry has grown from classical mechanics of the eighteenth and nineteenth centuries; see [10]. One of the most striking applications of symplectic geometry in representation theory is the orbit method; see [112].

A.3 The language of connections

The language of connections is indispensable in differential geometry and related areas. Here we provide a few essential formulæ and results, necessary for understanding the main body of the book. The reader is encouraged to consult the vast literature on the subject (e.g., [123, 193]).

Basic definitions

An *affine connection* on a smooth manifold M^n can be defined in terms of covariant differentiation, a bilinear operation on the space of vector fields on M, denoted by $\nabla_X Y$

and satisfying the identities:

$$\nabla_{fX}Y = f\nabla_X Y, \qquad \nabla_X(fY) = X(f)Y + f\nabla_X Y$$

for any smooth function f. If M is an affine space, one may take ∇ to be the directional derivative; this is a flat affine connection.

A curve $\gamma(t)$ is a geodesic if its velocity vector is parallel along the curve:

$$\nabla_{\gamma'(t)}\gamma'(t) = 0 \tag{A.10}$$

for all t. For a different parameterization $\gamma(\tau)$ of the same curve, one has:

$$\nabla_{\gamma'(\tau)}\gamma'(\tau) = \phi(\tau)\gamma'(\tau) \tag{A.11}$$

where ϕ is a function. Conversely, if (A.11) holds then one can change the parameterization so that (A.10) is satisfied.

The following tensors are of particular importance: the *torsion*

$$T(X, Y) = \nabla_X Y - \nabla_Y X - [X, Y],$$

the *curvature*

$$R(X, Y)Z = \nabla_X\nabla_Y Z - \nabla_Y\nabla_X Z - \nabla_{[X,Y]}Z,$$

and the *Ricci tensor*

$$\mathrm{Ric}(Y, Z) = \mathrm{Tr}\,(X \mapsto R(X, Y)Z).$$

We assume that the connections under consideration have zero torsion. A connection is called *flat* if $R = 0$.

Using the Leibnitz rule, one extends covariant differentiation to other tensor fields. For example, if g is a (pseudo)Riemannian metric then one defines $\nabla_X(g)$ by the equality:

$$(\nabla_X g)(Y, Z) = X(g(Y, Z)) - g(\nabla_X Y, Z) - g(Y, \nabla_X Z).$$

The *Levi–Civita connection* for a (pseudo)Riemannian metric g is a torsion-free connection preserving g; this connection is uniquely defined by g.

A connection ∇ is called *equiaffine* if it admits a parallel volume form Ω, namely, $\nabla_X\Omega = 0$ for all vector fields X.

Exercise A.3.1 A connection is equiaffine if and only if its Ricci tensor is symmetric: $\mathrm{Ric}(Y, Z) = \mathrm{Ric}(Z, Y)$.

Projective equivalence

Two torsion-free affine connections are called *projectively equivalent* if they have the same non-parameterized geodesics. The respective analytic condition is as follows.

Proposition A.3.2 *Two torsion-free connections $\widetilde{\nabla}$ and ∇ are projectively equivalent if and only if there exists a 1-form λ such that*

$$\widetilde{\nabla}_X Y = \nabla_X Y + \lambda(X)Y + \lambda(Y)X \tag{A.12}$$

for all vector fields X, Y.

Proof We only give an outline; see, for example, [156] for more detail.

Assume that (A.12) holds and $\gamma(t)$ is a geodesic of ∇. Then one has: $\widetilde{\nabla}_{\gamma'(t)}\gamma'(t) = \phi(t)\gamma'(t)$ with $\phi(t) = 2\lambda(\gamma'(t))$. It follows that γ is a geodesic of the connection $\widetilde{\nabla}$.

Conversely, let $\widetilde{\nabla}$ and ∇ be projectively equivalent. Let $K(X, Y) = \widetilde{\nabla}_X Y - \nabla_X Y$. Then K is a symmetric tensor. We claim that there exists a 1-form λ such that $K(X, X) = 2\lambda(X)X$ for any tangent vector X; the result will then follow by polarization.

Given X, let $\gamma(t)$ be the geodesic of ∇ such that $\gamma'(0) = X$. Then γ is also a geodesic of $\widetilde{\nabla}$, therefore, by (A.11), $K(X, X) = \widetilde{\nabla}_X X = \phi X$ where ϕ depends on X. It is a matter of linear algebra to prove that if $K(X, X) = \phi(X)X$ then ϕ is a linear function, and it remains to set $\lambda = \phi/2$. □

Exercise A.3.3 Let $\widetilde{\nabla}$ and ∇ be projectively equivalent equiaffine connections. Show that the 1-form λ in Proposition A.3.2 is closed.

Now we define a *projective connection* on a smooth manifold M. Similarly to many notions of differential geometry, this one is defined as an equivalence class of a special type of atlases on M. Namely, consider an open covering U_i such that each U_i carries an affine connection ∇_i, and the connections ∇_i and ∇_j are projectively equivalent on $U_i \cap U_j$. A projective connection is called *flat* if the connections ∇_i are projectively equivalent to flat affine connections. A flat projective connection is the same as a projective structure; see Section 6.1.

Example A.3.4 *Projective space $\mathbb{RP}^n$ has a flat projective connection whose geodesics are projective lines.*

One can prove that if M has a projective connection then there exists an affine connection ∇ on M such that, for all i, ∇ is projectively equivalent to ∇_i. If M has a volume form Ω then ∇ is uniquely defined by the condition that Ω is parallel.

Given an affine connection on M^n, the following tensor is called the *Weyl projective curvature*:

$$W(X, Y)Z = R(X, Y)Z - \frac{1}{n-1}(\mathrm{Ric}(Y, Z)X - \mathrm{Ric}(X, Z)Y).$$

Exercise A.3.5 Prove that the Weyl projective curvature tensor is invariant under projective equivalences of Ricci-symmetric affine connections.

A.4 The language of homological algebra

This section provides a very brief introduction to cohomology of Lie groups and Lie algebras; see, for example, [35, 72] for a comprehensive account.

Lie group and Lie algebra cohomology

Since Lie algebras are simpler than Lie groups, let us start with the former. Let $\mathfrak{g}$ be a Lie algebra and $T : \mathfrak{g} \to \mathrm{End}(V)$ its action on a space V. The space of q-dimensional cochains $C^q(\mathfrak{g}; V)$ of $\mathfrak{g}$ with coefficients in V consists of continuous skew-symmetric

q-linear function on $\mathfrak{g}$ with values in V. The differential $d_q : C^q(\mathfrak{g}; V) \to C^{q+1}(\mathfrak{g}; V)$ is given by the formula:

$$d_q\, c(X_1, \ldots, X_{q+1})$$
$$= \sum_{1 \le i < j \le q+1} (-1)^{i+j-1}\, c([X_i, X_j], X_1, \ldots, \hat{X}_i, \ldots, \hat{X}_j, \ldots, X_{q+1})$$
$$+ \sum_{1 \le i \le q+1} (-1)^i\, T(X_i)\, c(X_1, \ldots, \hat{X}_i, \ldots, X_{q+1}).$$

One has $d_q \circ d_{q-1} = 0$, and one defines the cohomology in the usual way: $H^q(\mathfrak{g}, V) = \operatorname{Ker} d_q / \operatorname{Im} d_{q-1}$.

The reader who has not previously encountered the above formula for the differential may find it quite formidable. The origin of this formula is in homological algebra. Note a remarkable similarity of this formula to the familiar formula for the exterior differential of a differential q-form on a smooth manifold.

Let G be a Lie group and $T : G \to \operatorname{End}(V)$ its action on a space V. The space of q-dimensional cochains $C^q(G, V)$ consists of smooth functions of q variables $G \times \cdots \times G \to V$, and the differential is given by the formula:

$$d_q\, C(g_1, \ldots, g_{q+1})$$
$$= T(g_1)\, C(g_2, \ldots, g_{q+1}) + \sum_{1 \le i \le q} (-1)^i\, C(g_1, \ldots, g_i g_{i+1}, \ldots, g_{q+1})$$
$$+ (-1)^{q+1}\, C(g_1, \ldots, g_q).$$

As before, one defines the cohomology $H^q(G, V) = \operatorname{Ker} d_q / \operatorname{Im} d_{q-1}$.

Of course, one may relax the differentiability assumption on cochains and define continuous, measurable, discrete, etc., cohomology (such a change may dramatically affect the result). We assumed throughout the book that the group cocycles are given by differentiable functions.

The cohomology of Lie algebras and Lie groups has relative versions which are defined via cochains that vanish once at least one argument belongs to the respective subalgebra or a subgroup.

One has a natural homomorphism $H^q(G; V) \to H^q(\mathfrak{g}; V)$, given, on the cocycle level, by the formula

$$c(X_1, \ldots, X_q)$$
$$= \frac{\mathrm{d}}{\mathrm{d}t}\, C\left(\exp tX_1, \exp(-tX_1)\exp tX_2, \ldots, \exp(-tX_{q-1})\exp tX_q\right)\Big|_{t=0}.$$

The inverse problem is to determine whether a cocycle on a Lie algebra $\mathfrak{g}$ corresponds to some cocycle on the respective Lie group G. This is the problem of *integration* of algebra cocycles to group cocycles.

Exercise A.4.1 Check that:
(a) $d_q \circ d_{q-1} = 0$, both for Lie groups and Lie algebras;
(b) the above homomorphism $C^q(G; V) \to C^q(\mathfrak{g}; V)$ commutes with the differentials.

The cohomology of Lie algebras and Lie groups, especially the first and second cohomology, has numerous applications and interpretations. We discuss two such interpretations, most pertinent to this book, in the next two subsections.

Affine modules and first cohomology

Let G be a group, V a G-module and $T : G \to \mathrm{End}(V)$ the G-action on V. A structure of an affine module on V is a structure of a G-module on the space $V \oplus \mathbb{R}$ defined by

$$\widetilde{T}_g : (v, \alpha) \mapsto (T_g\, v + \alpha\, C(g), \alpha), \tag{A.13}$$

where C is a map from G to V. The condition $\widetilde{T}_g \circ \widetilde{T}_h = \widetilde{T}_{gh}$ is equivalent to

$$C(gh) = T_g\, C(h) + C(g).$$

Therefore C is a 1-cocycle on G with coefficients in V. The module (A.13) is also called an *extension* of V.

The equivalence relation is defined by the following commutative diagram in which B is an isomorphism of G-modules:

$$\begin{array}{ccccccccc} 0 & \longrightarrow & V & \longrightarrow & V \oplus \mathbb{R} & \longrightarrow & \mathbb{R} & \longrightarrow & 0 \\ & & \downarrow{\scriptstyle \mathrm{Id}} & & \downarrow{\scriptstyle B} & & \downarrow{\scriptstyle \mathrm{Id}} & & \\ 0 & \longrightarrow & V & \longrightarrow & V \oplus \mathbb{R} & \longrightarrow & \mathbb{R} & \longrightarrow & 0 \end{array}$$

The simplest extension corresponds to $C = 0$; the extension (A.13) is called trivial if it is equivalent to this simplest one. This corresponds to the fact that the cocycle c is of a special form:

$$C(g) = T_g\, v - v \tag{A.14}$$

with some fixed $v \in V$, that is, C is a coboundary.

Thus the space $H^1(G; V)$ is isomorphic to the space of equivalence classes of extensions of module V.

Central extensions

Let $\mathfrak{g}$ be a Lie algebra. A central extension of $\mathfrak{g}$ is a Lie algebra structure on the space $\mathfrak{g} \oplus \mathbb{R}$ defined by the commutator

$$[(X, \alpha), (Y, \beta)] = ([X, Y], c(X, Y))$$

where $X, Y \in \mathfrak{g},\ \alpha, \beta \in \mathbb{R}$ and c is a skew-symmetric bilinear map from $\mathfrak{g}$ to $\mathbb{R}$. The Jacobi identity for the above commutator is equivalent to the following relation:

$$c(X, [Y, Z]) + c(Y, [Z, X]) + c(Z, [X, Y]) = 0. \tag{A.15}$$

Therefore c is a 2-cocycle. In other words, a central extension is defined by an exact sequence of Lie algebras

$$0 \longrightarrow \mathbb{R} \longrightarrow \mathfrak{g} \oplus \mathbb{R} \longrightarrow \mathfrak{g} \longrightarrow 0. \tag{A.16}$$

The simplest extension corresponds to $c = 0$; the extension (A.16) is called trivial (or a direct sum) if it is isomorphic to this simplest one. This corresponds to the fact that the cocycle c is a coboundary:

$$c(X, Y) = \phi([X, Y])$$

where ϕ is a linear function.

Therefore the space $H^2(\mathfrak{g}; \mathbb{R})$ is isomorphic to the space of equivalence classes of central extensions of the Lie algebra $\mathfrak{g}$. Note that the space $H^2(\mathfrak{g}; V)$, where V is a $\mathfrak{g}$-module, has a similar interpretation.

A.5 Remarkable cocycles on groups of diffeomorphisms

This section presents a number of cocycles on groups of diffeomorphisms given by explicit geometrical constructions. There are only a few known constructions which have a tendency to reappear in various areas. Some of these constructions were used in this book, starting with the Schwarzian derivative. As we mentioned earlier, the problem of integration of Lie algebra cocycles to Lie group cocycles is highly non-trivial, and this section will provide examples of such integration.

A 2-cocycle on SL(2, $\mathbb{R}$)

The cohomology of finite-dimensional classical Lie groups with trivial coefficients is described by the following isomorphism:

$$H^q(G; \mathbb{R}) = H^q_{\mathrm{top}}(\bar{G}/K)$$

where H_{top} means the cohomology of a topological space, $\bar{G}$ is a compact form of the group G and K is the maximal compact subgroup of G. The compact form is a compact Lie group such that the complexification of $\bar{\mathfrak{g}}$ coincides with that of $\mathfrak{g}$ (see, e.g., [72]).

If $G = \mathrm{SL}(2, \mathbb{R})$ then $\bar{G} = \mathrm{SU}(2)$ and $K = \mathrm{SO}(2)$, hence $\bar{G}/K = S^2$. Therefore $H^q(\mathrm{SL}(2, \mathbb{R}); \mathbb{R}) = H^q(S^2)$ so that the only non-trivial cohomology of SL(2, $\mathbb{R}$) is the second one. Let us construct a corresponding 2-cocycle.

Recall that PGL(2, $\mathbb{R}$) can be identified with the group of orientation preserving isometries of the hyperbolic plane, considered in the upper half-plane model. Thus one has an action of SL(2, $\mathbb{R}$) in the hyperbolic plane by isometries. Pick a point x and define

$$C(f, g) = \text{Area}\,(x, f(x), fg(x)), \tag{A.17}$$

the signed area of a triangle in the hyperbolic plane.

Exercise A.5.1 Check that $C(f, g)$ is a 2-cocycle.

It could well be that the constructed cocycle is trivial; however, this is not the case.

Lemma A.5.2 *The cocycle $C(f, g)$ is non-trivial.*

Proof If C is trivial then there exists a function $B : \mathrm{SL}(2, \mathbb{R}) \to \mathbb{R}$ such that

$$B(fg) - B(f) - B(g) = C(f, g). \tag{A.18}$$

It is clear from (A.17) that $C(f, g) = 0$ if f is identity or if $g = f^{-1}$. It follows from (A.18) that $B(Id) = 0$ and $B(f^{-1}) = -B(f)$.

Consider the group H, the stabilizer of point x. For $f \in H$ one has, by (A.17), that $C(f, g) = 0$. It follows from (A.18) that $B : H \to \mathbb{R}$ is a homomorphism. Since $H = S^1$, this homomorphism is trivial.

Assume now that $f(x) = g(x)$. Then $f^{-1}g \in H$ and hence $B(f^{-1}g) = 0$. It also follows from (A.17) that $C(f, g) = 0$, and then (A.18) implies that $B(f) = B(g)$. Therefore the value of $B(f)$ depends only on the point $f(x)$, that is, $B(f) = F(f(x))$ for some function F.

Finally assume that $F(y) \neq 0$ for some point y. Let f be the reflection in the middle of the segment xy. Then $f^{-1} = f$ and $f(x) = y$. It follows that $F(y) = B(f^{-1}) = -B(f) = -F(y)$. Hence $F(y) = 0$, a contradiction. □

Bott–Thurston cocycle

Recall the Gelfand–Fuchs cocycle (1.31) of the Lie algebra $\mathrm{Vect}(S^1)$, "responsible" for the Virasoro algebra, which we rewrite here as

$$c\left(h_1(x)\frac{\mathrm{d}}{\mathrm{d}x}, h_2(x)\frac{\mathrm{d}}{\mathrm{d}x}\right) = \int_{S^1} \begin{vmatrix} h_1'(x) & h_2'(x) \\ h_1''(x) & h_2''(x) \end{vmatrix} \mathrm{d}x. \tag{A.19}$$

The computation of the ring $H^*(\mathrm{Vect}(S^1); \mathbb{R})$ was one of the first results on cohomology of infinite-dimensional Lie algebras, obtained by Gelfand and Fuchs in 1968. The result of this computation is as follows: $H^*(\mathrm{Vect}(S^1); \mathbb{R})$ is the tensor product of the polynomial ring with one two-dimensional generator and the exterior algebra with one three-dimensional generator. The two-dimensional generator is represented by the cocycle (A.19) and the three-dimensional generator by the cocycle

$$\bar{c}\left(h_1(x)\frac{\mathrm{d}}{\mathrm{d}x}, h_2(x)\frac{\mathrm{d}}{\mathrm{d}x}, h_3(x)\frac{\mathrm{d}}{\mathrm{d}x}\right) = \int_{S^1} \begin{vmatrix} h_1(x) & h_2(x) & h_3(x) \\ h_1'(x) & h_2'(x) & h_3'(x) \\ h_1''(x) & h_2''(x) & h_3''(x) \end{vmatrix} \mathrm{d}x. \tag{A.20}$$

We prove here that the Gelfand–Fuchs 2-cocycle is non-trivial. The same fact for the above 3-cocycle will become clear in Appendix A.6.

Lemma A.5.3 *The Gelfand–Fuchs cocycle (A.19) is non-trivial.*

Proof The following identity holds in the Lie algebra $\mathrm{Vect}(S^1)$ but fails in the Virasoro algebra (see Section 1.6):

$$\mathrm{Alt}_{1,2,3,4}\,[X_1, [X_2, [X_3, [X_4, Y]]]] = 0. \tag{A.21}$$

Exercise A.5.4 (a) Prove the identity (A.21) in $\mathrm{Vect}(S^1)$;

(b) Prove that the $\mathbb{R}$-component of the expression (A.21) in the Virasoro algebra equals

$$\int_{S^1} \begin{vmatrix} h_1(x) & h_2(x) & h_3(x) & h_4(x) \\ h_1'(x) & h_2'(x) & h_3'(x) & h_4'(x) \\ h_1''(x) & h_2''(x) & h_3''(x) & h_4''(x) \\ h_1'''(x) & h_2'''(x) & h_3'''(x) & h_4'''(x) \end{vmatrix} g(x)\, \mathrm{d}x.$$

where $X_i = h_i(x)\,\mathrm{d}/\mathrm{d}x$ and $Y = g(x)\,\mathrm{d}/\mathrm{d}x$.

Thus the Virasoro algebra is not isomorphic to $\mathrm{Vect}(S^1) \oplus \mathbb{R}$ and the cocycle (A.19) is non-trivial. □

The cohomology ring of the respective Lie group $\mathrm{Diff}_+(S^1)$ has the following description (see [72]): it has two generators, $\alpha, \beta \in H^2(\mathrm{Diff}_+(S^1); \mathbb{R})$ with the only relation $\beta^2 = 0$. The cohomology class α is taken by the natural homomorphism $H^2(\mathrm{Diff}_+(S^1)) \to H^2(\mathrm{Vect}(S^1))$ to the generator of $H^2(\mathrm{Vect}(S^1))$, represented by the cocycle c given by (A.19). The problem of integrating this Lie algebra cocycle to the group $\mathrm{Diff}_+(S^1)$ was solved by Thurston and Bott; see [31].

Let us start with a simple general construction. Let M be a smooth manifold with a volume form μ. Assign to an orientation preserving diffeomorphism $f : M \to M$ a smooth function $\bar{C}(f)$ on M defined by the formula:

$$f^*\mu = e^{\bar{C}(f)}\mu.$$

Set $C(f) = \bar{C}(f^{-1})$. One obtains a map $C : \mathrm{Diff}_+(M) \to C^\infty(M)$.

Exercise A.5.5 Check that C is a 1-cocycle.

Another piece of "general nonsense" contributing to the Bott–Thurston construction is the next recipe for making 2-cocycles from 1-cocycles. Let G be a Lie group acting on spaces U and V, $\phi : \wedge^2 U \to V$ a G-homomorphism and $C \in C^1(G; U)$ a 1-cocycle.

Exercise A.5.6 Check that the formula $D(g_1, g_2) = \phi(C(g_1), C(g_1 g_2))$ defines a 2-cocycle of G with coefficients in V.

Back to the group $\mathrm{Diff}_+(S^1)$. For $M = S^1$, one takes $\mu = \mathrm{d}x$ and obtains a 1-cocycle $C(f)$ as in Exercise A.5.5. For a $\mathrm{Diff}_+(S^1)$-invariant pairing, take

$$\phi : \wedge^2 C^\infty(S^1) \to \mathbb{R}, \qquad \phi(f, g) = \int_{S^1} f\, \mathrm{d}g.$$

Then the cocycle of Exercise A.5.6 becomes the Bott–Thurston cocycle in $C^2(\mathrm{Diff}_+(S^1); \mathbb{R})$ given by the formula

$$D(f, g) = \int_{S^1} C(f)\, \mathrm{d}C(fg).$$

A projectively invariant cocycle representing the same cohomology class was given in [53].

Cocycles as coboundaries of "ghosts"

The relation between affine modules and the first cohomology discussed in Appendix A.4 can provide a source of 1-cocycles of a Lie group or a Lie algebra. Let A be an affine space with the underlining vector space V, acted upon by a Lie group G; this means that G acts on A by affine transformations. Fix $a \in A$ and consider a map $G \to V$ given by the formula

$$C(g) = g(a) - a. \tag{A.22}$$

This formula is identical to that for a coboundary (A.14), in particular, $C(g)$ is a 1-cocycle of G with coefficients in V. However, this cocycle does not have to be trivial, since a is not an element of V. In this sense, $C(g)$ is the coboundary of a "ghost".

Exercise A.5.7 Show that replacing the point $a \in A$ in (A.22) by another point changes $C(g)$ by a coboundary.

An example of this construction is the multi-dimensional Schwarzian in Section 7.1: the affine space is that of projective connections, the respective vector space is the space of tensor-valued 1-forms, and the group consists of diffeomorphisms of the manifold.

Another example is the space of projective structures on S^1, as in Section 1.3: the affine space consists of Sturm–Liouville operators, the respective vector space is the space of quadratic differentials and the group is Diff(S^1); see (1.12) for the action on the potential of a Sturm–Liouville operator. The resulting cocycle is the Schwarzian derivative.

A.6 The Godbillon–Vey class

We feel that it would be an unforgivable omission not to mention the beautiful Godbillon–Vey construction of characteristic classes of codimension 1 foliations.

Construction

A smooth k-dimensional foliation on a smooth manifold M^n is a partition of M into connected subsets (the leaves), locally diffeomorphic to the partition of $\mathbb{R}^n$ into parallel k-dimensional planes. A codimension 1 foliation $\mathcal{F}$ is an integrable codimension 1 distribution, locally given by a 1-form α, satisfying the Frobenius integrability condition $\alpha \wedge \mathrm{d}\alpha = 0$. Assume that the foliation is coorientable, so that the form α is globally defined.

The Frobenius integrability condition is equivalent to the existence of a 1-form η such that $\mathrm{d}\alpha = \eta \wedge \alpha$. Consider the 3-form $\eta \wedge \mathrm{d}\eta$.

Exercise A.6.1 Check that $\mathrm{d}\eta = \nu \wedge \alpha$ for some 1-form ν, the 3-form $\eta \wedge \mathrm{d}\eta$ is closed, and that its de Rham cohomology class in $H^3(M, \mathbb{R})$ does not depend on the choices involved.

Hint. What happens when α is replaced by $f\alpha$ with f a non-vanishing function or η by $\eta + g\alpha$ with g a function?

The cohomology class of $\eta \wedge \mathrm{d}\eta$ is the Godbillon–Vey class of the foliation $\mathcal{F}$, denoted by $\mathrm{gv}(\mathcal{F})$. On an orientable closed three-dimensional manifold M, one also defines the Godbillon–Vey number $\int_M \mathrm{gv}(\mathcal{F})$.

Non-triviality

To construct a foliation with non-trivial Godbillon–Vey class, consider the group $G = \mathrm{SL}(2, \mathbb{R})$. Let $H \subset G$ be the two-dimensional subgroup, consisting of the matrices with zero in the left lower corner. Let $\Gamma \subset G$ be a discrete subgroup with compact quotient space G/Γ. The right cosets of H foliate G, and this foliation is invariant under the left action by Γ. Therefore one obtains a codimension 1 foliation $\mathcal{F}(F, H, \Gamma)$ on the manifold G/Γ. This foliation is an example of a *homogeneous foliation* (another familiar example being a linear foliation on the torus).

Let $\omega_{-1}, \omega_0, \omega_1$ be the right-invariant 1-forms on G given by the formulae

$$\omega_{-1}\begin{pmatrix} x & y \\ z & -x \end{pmatrix} = z, \quad \omega_0\begin{pmatrix} x & y \\ z & -x \end{pmatrix} = x, \quad \omega_1\begin{pmatrix} x & y \\ z & -x \end{pmatrix} = y.$$

Then

$$\mathrm{d}\omega_{-1} = 2\omega_{-1} \wedge \omega_0, \quad \mathrm{d}\omega_0 = \omega_{-1} \wedge \omega_1, \quad \mathrm{d}\omega_1 = 2\omega_0 \wedge \omega_1 \tag{A.23}$$

and $\omega_{-1} \wedge \omega_0 \wedge \omega_1$ is a bi-invariant volume form. The forms ω_i are pull-backs of 1-forms $\bar{\omega}_i$ on G/Γ satisfying the same relations (A.23) therein. The foliation $\mathcal{F}(F, H, \Gamma)$ is given by the 1-form $\bar{\omega}_{-1}$, and a calculation using (A.23) yields: $\mathrm{gv}(\mathcal{F}(F, H, \Gamma)) = -4\bar{\omega}_{-1} \wedge \bar{\omega}_0 \wedge \bar{\omega}_1$.

Remark A.6.2 Using the fact that $\mathrm{SL}(2, \mathbb{R})$ transitively acts on the hyperbolic plane by isometries, one can choose Γ in such a way that G/Γ identifies with the unit tangent bundle of a surface of genus ≥ 2. Then $\mathcal{F}(F, H, \Gamma)$ is the horocycle foliation.

Relation to the cohomology of infinite-dimensional Lie algebras

What is the relation, if any, between this construction and Lie algebra cohomology? To understand this relation, introduce another infinite-dimensional Lie algebra W_1, the algebra of formal vector fields on the line. Its elements are $h(x)\,\mathrm{d}/\mathrm{d}x$ where $h(x)$ is a formal power series. The only non-trivial cohomologies of W_1 (with trivial coefficients) are $H^0(W_1; \mathbb{R}) = H^3(W_1; \mathbb{R}) = \mathbb{R}$. A non-trivial 3-cocycle is given by the formula

$$\tilde{c}\left(h_1(x)\frac{\mathrm{d}}{\mathrm{d}x}, h_2(x)\frac{\mathrm{d}}{\mathrm{d}x}, h_3(x)\frac{\mathrm{d}}{\mathrm{d}x}\right) = \begin{vmatrix} h_1(0) & h_2(0) & h_3(0) \\ h_1'(0) & h_2'(0) & h_3'(0) \\ h_1''(0) & h_2''(0) & h_3''(0) \end{vmatrix}. \tag{A.24}$$

One has a homomorphism $\mathrm{Vect}(S^1) \to W_1$, given by taking the infinite jet of a vector field at an arbitrarily chosen point of S^1. The pull-back of the cocycle (A.24) is cohomologous to the cocycle (A.20).

Let $\mathcal{F}$ be a coorientable codimension 1 foliation on M. We will construct a 1-form ω on M with values in the algebra W_1. This form will satisfy the Maurer–Cartan equation

$$\mathrm{d}\omega = -\frac{1}{2}[\omega, \omega]. \tag{A.25}$$

Fix a smooth map $M \times \mathbb{R} \to M$ that, for each $x \in M$, sends $x \times \mathbb{R}$ to a small curve, transverse to $\mathcal{F}$ at point x, sending $(x, 0)$ to x and compatible with the coorientation. Let $x \in M$ be a point and $v \in T_x M$ a tangent vector. Include v into a vector field in a vicinity of x and let ϕ_t be the respective local one-parameter group of diffeomorphisms. Let $y = \phi_\varepsilon(x)$. The intersections of the transverse curves with the leaves of the foliation define the (germ of the) holonomy diffeomorphism $\psi_\varepsilon : \mathbb{R} \to \mathbb{R}$. Note that ψ_0 is the identity. Taking the derivative yields a linear map

$$\omega : v \mapsto \frac{\mathrm{d}\psi_\varepsilon}{\mathrm{d}\varepsilon}|_{\varepsilon=0}.$$

This is the desired 1-form $\omega : T_x M \to W_1$.

Exercise A.6.3 Let the foliation be given by a 1-form α such that $\mathrm{d}\alpha = \alpha \wedge \eta$ and $\mathrm{d}\eta = \nu \wedge \alpha$. Check that

$$\omega(v) = \alpha(v)\frac{\mathrm{d}}{\mathrm{d}x} + \eta(v)\, x\frac{\mathrm{d}}{\mathrm{d}x} + \nu(v)\, x^2\frac{\mathrm{d}}{\mathrm{d}x} + \cdots.$$

The 1-form ω induces a homomorphism $H^*(W_1; \mathbb{R}) \to H^*(M, \mathbb{R})$. Namely, let c be a q-cocycle on W_1; define a q-form on M by the formula:

$$\omega_c(v_1, \ldots, v_q) = c(\omega(v_1), \ldots, \omega(v_q)).$$

Exercise A.6.4 Using the Maurer–Cartan formula (A.25), check that the map $c \mapsto \omega_c$ commutes with differentials and deduce that ω_c is a closed differential form.

Taking the 3-cocycle (A.24) in this construction yields the Godbillon–Vey class $\mathrm{gv}(\mathcal{F})$.

Remark A.6.5 The 1-form ω is an example W_1-structure on M. Let $\mathfrak{g}$ be a Lie algebra. A $\mathfrak{g}$-structure on a smooth manifold M is a $\mathfrak{g}$-valued differential 1-form satisfying the Maurer–Cartan equation (A.25). A codimension q foliation on M with trivialized normal bundle determines a W_q-structure where W_q is the Lie algebra of formal vector fields in $\mathbb{R}^q$. Characteristic classes of codimension q foliations correspond to $H^*(W_q; \mathbb{R})$.

Comment

Godbillon and Vey discovered the invariant $\mathrm{gv}(\mathcal{F})$ in 1971. In spite of the simplicity of the definition, the geometric meaning of this class remains somewhat mysterious;[1] see [80, 178] for detailed accounts. Unlike the usual characteristic classes (such as Chern or Stifel–Whitney classes), the Godbillon–Vey number can vary in one-parameter families;

[1] The reader might want to muse on Thurston's description of the Godbillon–Vey invariant as the "helical wobble"; see [219].

this phenomenon was discovered by Thurston. The theory of characteristic classes of foliations is one of the most impressive applications of Gelfand–Fuchs cohomology; see [72] and [81, 214] for the theory of foliations.

A.7 The Adler–Gelfand–Dickey bracket and infinite-dimensional Poisson geometry

In Section 1.6 we encountered the Virasoro algebra and its connection with Sturm–Liouville operators was revealed. What about higher-order differential operators? In this section we define a remarkable infinite-dimensional Poisson structure on the space of differential operators (2.7), called the Adler–Gelfand–Dickey bracket. This algebraic structure generalizes the Virasoro algebra and plays a similar role in the theory of integrable systems and conformal fields theory.

We classify symplectic leaves of the Adler–Gelfand–Dickey bracket and show that this classification problem is equivalent to the classification of non-degenerate projective curves discussed in Section 2.3. This provides a geometric interpretation of this complicated infinite-dimensional Poisson structure.

Introducing the Adler–Gelfand–Dickey bracket

The Adler–Gelfand–Dickey bracket is defined on the space of linear differential operators on S^1

$$A = \frac{\mathrm{d}^{n+1}}{\mathrm{d}x^{n+1}} + a_{n-1}(x)\frac{\mathrm{d}^{n-1}}{\mathrm{d}x^{n-1}} + \cdots + a_1(x)\frac{\mathrm{d}}{\mathrm{d}x} + a_0(x), \tag{A.26}$$

already familiar to the reader. This is an affine space and we identify all cotangent spaces by parallel translation. Similarly to Section 1.6, we consider the space of regular functionals of the form

$$\varphi(A) = \sum_{i=0}^{n-1} \int_{S^1} b_i(x)a_i(x)\mathrm{d}x \tag{A.27}$$

where $b_i \in C^\infty(S^1)$.

It is convenient to represent such functionals by (formal) pseudo-differential operators

$$P = \sum_{k=1}^{\infty} c_k(x)\left(\frac{\mathrm{d}}{\mathrm{d}x}\right)^{-k}. \tag{A.28}$$

For such operators the *Adler trace* is defined by

$$\mathrm{Tr}(P) = \int_{S^1} c_{-1}(x)\mathrm{d}x.$$

The pairing between an operator and a pseudo-differential operator is defined by the formula

$$\langle P, A\rangle = \mathrm{Tr}(P \circ A).$$

We denote this functional by φ_P.

Remark A.7.1 One deals with pseudo-differential operators similarly to differential operators; in particular, one composes them using the following rule:

$$\left(\frac{\mathrm{d}}{\mathrm{d}x}\right)^{-1} \circ a(x) = a(x)\left(\frac{\mathrm{d}}{\mathrm{d}x}\right)^{-1} + \sum_{i=1}^{\infty}(-1)^i a^{(i)}(x)\left(\frac{\mathrm{d}}{\mathrm{d}x}\right)^{-i-1}$$

which amounts to integration by parts. In this way, one rewrites the composition $P \circ A$ in the form (A.28).

Exercise A.7.2 The functional (A.27) can be written in the form $\langle P, A\rangle$ where

$$P = \left(\frac{\mathrm{d}}{\mathrm{d}x}\right)^{-n} b_{n-1}(x) + \cdots + \left(\frac{\mathrm{d}}{\mathrm{d}x}\right)^{-1} b_0(x).$$

Hint. Integral of the derivative of a function over S^1 equals zero.

To define the Adler–Gelfand–Dickey bracket, it suffices to define it for linear functionals:

$$\{\varphi_P, \varphi_Q\}(A) = \mathrm{Tr}\left((A \circ Q)_+ \circ A \circ P - P \circ A \circ (Q \circ A)_+\right) \tag{A.29}$$

where the subscript $+$ stands for cutting off the negative powers of $\mathrm{d}/\mathrm{d}x$. According to a Gelfand and Dickey theorem, formula (A.29) defines a Poisson structure.

Model example: third-order operators

Let us consider in more detail an instructive particular case $n = 2$. The differential operators under consideration are

$$A = \frac{\mathrm{d}^3}{\mathrm{d}x^3} + a_1(x)\frac{\mathrm{d}}{\mathrm{d}x} + a_0(x). \tag{A.30}$$

In the space of functionals (A.27), it is natural to choose the basis:

$$\varphi_h(A) = \int_{S^1} h(x)a_1(x)\mathrm{d}x, \qquad \psi_f(A) = \int_{S^1} f(x)\left(a_0(x) - \frac{a_1'(x)}{2}\right)\mathrm{d}x.$$

Note that a_1 has the meaning of a quadratic differential while $a_0 - a_1'/2$ of a cubic form on a projective curve; see Section 1.4. It follows that h has the meaning of a vector field while f of a tensor density of degree -2.

Exercise A.7.3 (a) Show that the functionals φ_h form the Virasoro algebra:

$$\{\varphi_{h_1}, \varphi_{h_2}\} = \varphi_{h_1h_2'-h_1'h_2} + 4\int_{S^1} h_1h_2'''\mathrm{d}x.$$

(b) Show that the Poisson bracket between the functionals φ_h and ψ_f corresponds to the action of the vector field $h(x)\mathrm{d}/\mathrm{d}x$ on the density $f(x)(\mathrm{d}x)^{-2}$:

$$\{\varphi_h, \psi_f\} = \psi_{hf'-2h'f}.$$

(c)* Compute the Poisson bracket between ψ_{f_1} and ψ_{f_2}, a complicated quadratic functional.

Our next goal is to compute the Hamiltonian vector fields X_{φ_h} and X_{ψ_f} corresponding to the functionals φ_h and ψ_f.

Exercise A.7.4 Prove that the Hamiltonian vector field X_{φ_h} is precisely the action of the vector field $h(x)\mathrm{d}/\mathrm{d}x$ on the space of differential operators:

$$X_{\varphi_h} = L^2_{h(x)\mathrm{d}/\mathrm{d}x} \circ A - A \circ L^{-1}_{h(x)\mathrm{d}/\mathrm{d}x}$$

where

$$L^{\lambda}_{h(x)\mathrm{d}/\mathrm{d}x} = h(x)\mathrm{d}/\mathrm{d}x + \lambda h'(x)$$

is the Lie derivative (1.28) on the space of λ-densities. This formula is the infinitesimal version of the $\mathrm{Diff}(S^1)$-action (2.9).

The explicit formula for the Hamiltonian vector field X_{ψ_f} is complicated. We compute X_{ψ_f} in terms of solutions of the equation $Ay = 0$. The action on operators and solutions are related by

$$X_{\psi_f}(A)\, y + A\, X_{\psi_f}(y) = 0.$$

Exercise A.7.5 (a) Show that the Hamiltonian vector field X_{ψ_f} is defined by the formula

$$X_{\psi_f}(y) = \tfrac{1}{6}\, J_2^{-2,-1}(f, y) + \tfrac{2}{3}\, f y\, a_1 \qquad \text{(A.31)}$$

where $J_2^{-2,-1}$ is the second-order transvectant (3.1), namely,

$$J_2^{-2,-1}(f, y) = 6 f y'' - 3 f' y' + f'' y.$$

(b) Show that the action of X_{φ_h} on solutions becomes the Lie bracket of vector fields:

$$X_{\varphi_h}(y) = h y' - h' y. \qquad \text{(A.32)}$$

Remark A.7.6 The appearance of the transvectants in the context of the Adler–Gelfand–Dickey bracket is remarkable. We see how projective differential geometry intervenes, sometimes unexpectedly, into various geometric issues. Whenever it happens, it clearly indicates the right way of thinking about a problem.

Symplectic leaves

Recall the correspondence between the differential operators (A.26) and non-degenerate curves in $\mathbb{RP}^n$; see Section 2.2. Although the coefficients of an operator are periodic functions, the corresponding curve is not necessarily closed. Instead it satisfies the monodromy condition

$$\gamma(x + 2\pi) = \mathrm{T}(\gamma(x)), \qquad \Gamma(x + 2\pi) = \mathrm{T}(\Gamma(x)),$$

where T is a representative of a conjugacy class in $\mathrm{SL}(n+1, \mathbb{R})$.

Theorem A.7.7 *Symplectic leaves of the Adler–Gelfand–Dickey bracket are in one-to-one correspondence with the homotopy classes of non-degenerate curves with fixed monodromy.*

Proof We prove this theorem in the particular case $n = 2$ discussed above. The proof in the general case is similar.

To start with, the action (A.31)–(A.32) preserves the monodromy simply because it is linear in y. Let us show that monodromy is the unique local invariant of symplectic leaves.

As in Section 1.6, let us use the homotopy method. Consider a family A_t of differential operators (A.30) with fixed monodromy. We want to prove that there exist h_t and f_t such that, for every t,

$$\dot{A}_t = X_{\varphi_{h_t}}(A) + X_{\psi_{f_t}}(A). \tag{A.33}$$

To solve this homotopy equation, fix a basis $(y_{1_t}(x), y_{2_t}(x), y_{3_t}(x))$ of solutions to the equation $A_t y_t = 0$ such that the two conditions hold: (a) the monodromy matrix does not depend on t, (b) the Wronski determinant

$$\begin{vmatrix} y_{1_t} & y_{2_t} & y_{3_t} \\ y_{1_t}' & y_{2_t}' & y_{3_t}' \\ y_{1_t}'' & y_{2_t}'' & y_{3_t}'' \end{vmatrix} \equiv 1.$$

Exercise A.7.8 Prove that the formula

$$h = -\frac{1}{2}\begin{vmatrix} y_{1_t} & y_{2_t} & y_{3_t} \\ y_{1_t}' & y_{2_t}' & y_{3_t}' \\ \dot{y}_{1_t}' & \dot{y}_{2_t}' & \dot{y}_{3_t}' \end{vmatrix} + \frac{1}{2}\begin{vmatrix} y_{1_t} & y_{2_t} & y_{3_t} \\ y_{1_t}'' & y_{2_t}'' & y_{3_t}'' \\ \dot{y}_{1_t} & \dot{y}_{2_t} & \dot{y}_{3_t} \end{vmatrix}, \qquad f = \begin{vmatrix} y_{1_t} & y_{2_t} & y_{3_t} \\ y_{1_t}' & y_{2_t}' & y_{3_t}' \\ \dot{y}_{1_t} & \dot{y}_{2_t} & \dot{y}_{3_t} \end{vmatrix}$$

solves the homotopy equation (A.33).

Hint. Use the same idea as in the proof of Theorem 1.6.4.

We proved that for any smooth family A_t there exists a couple (h_t, f_t) such that the tangent vector $\dot{A}_t$ is given by the Hamiltonian vector field $X_{\varphi_{h_t}} + X_{\psi_{f_t}}$. By definition of symplectic leaves, this means that the family A_t belongs to the same symplectic leaf. □

Comment

The Adler–Gelfand–Dickey bracket appeared in [77, 78] and in [1]. The main interest in this structure is in the theory of completely integrable systems. Although the Adler–Gelfand–Dickey bracket is quadratic, it generalizes the commutator of the Virasoro algebra; cf. Exercise A.7.3. The Adler–Gelfand–Dickey structure is also known as the second Poisson structure for the generalized KdV equations; there is another related Poisson structure, known as the first one. Together these two structures form a Poisson pair; see [51, 49].

These structures play an important role in conformal field theory. In the physics literature, one usually chooses the Fourier basis in the space of functionals (A.27) and obtains a set of generators of an infinite-dimensional Lie algebra with quadratic relations. These algebras are called the classical W-algebras.

The relation between symplectic leaves of the Adler–Gelfand–Dickey bracket with non-degenerate curves in $\mathbb{RP}^n$ was established in [161, 106]. Our proof of Theorem A.7.7 in the case $n = 2$ is borrowed from [161], the general case was settled in [124]. Relations to transvectants are discussed in [168].

References

[1] M. Adler, On a trace functional for formal pseudo differential operators and the symplectic structure of the Korteweg-de Vries type equations, *Invent. Math.* **50** (1978/79), 219–248

[2] M. Akivis and V. Goldberg, *Projective Differential Geometry of Submanifolds*, North-Holland, 1993

[3] J.-C. Alvarez, Hilbert's fourth problem in two dimensions, *MASS Selecta*, American Mathematical Society, 2003, 165–184

[4] J.-C. Alvarez, Symplectic geometry and Hilbert's fourth problem, preprint

[5] J.-C. Alvarez, I. Gelfand and M. Smirnov, *Crofton Densities, Symplectic Geometry and Hilbert's Fourth Problem*, Arnold-Gelfand Mathematical Seminars, Birkhäuser, 1997, 77–92

[6] S. Angenent, Inflection points, extatic points and curve shortening, *Hamiltonian Systems with Three or More Degrees of Freedom*, Kluwer, 1999, 3–10

[7] V. Arnold, Sturm theory and symplectic geometry, *Func. Anal. Appl.* **19**:4 (1985), 1–10

[8] V. Arnold, The branched covering $\mathbb{CP}^2 \to S^4$, hyperbolicity and projective topology, *Sib. Math. J.* **29** (1988), 717–726

[9] V. Arnold, *Geometrical Methods in the Theory of Ordinary Differential Equations*, Springer-Verlag, 1988

[10] V. Arnold, *Mathematical Methods of Classical Mechanics*, Springer-Verlag, 1989

[11] V. Arnold, *Topological Invariants of Plane Curves and Caustics*, University Lecture Series 5, American Mathematical Society, 1994

[12] V. Arnold, Topological problems of the theory of wave propagation, *Russ. Math. Surv.* **51**:1 (1996), 1–47

[13] V. Arnold, *On the Number of Flattening Points on Space Curves*, American Mathematical Society Translations, ser. 2, vol. 171, 1996, 11–22

[14] V. Arnold, *Remarks on the Extactic Points of Plane Curves*, The Gelfand Mathematical Seminars, Birkhäuser, 1996, 11–22

[15] V. Arnold and A. Givental, Symplectic geometry, in *Encyclopedia of Mathematical Sciences, Dynamical Systems 4*, Springer-Verlag, 1990, 1–136

[16] V. Arnold, S. Gusein-Zade and A. Varchenko, *Singularities of differentiable maps.* vol. I. *The Classification of Critical Points, Caustics and Wave Fronts, Birkhäuser*, 1985

[17] D. Bao, S.-S. Chern and Z. Shen, *An Introduction to Riemann–Finsler Geometry*, Springer-Verlag, 2000

[18] M. Barner, Uber die Mindestanzahl stationarer Schmiegebenen bei geschlossenen strengconvexen Raumcurven, *Abh. Math. Sem. Univ. Hamburg* **20** (1956), 196–215

[19] L. Bates, A weak counterexample to the Carathéodory conjecture, *Diff. Geom. Appl.* **15** (2001), 79–80

[20] P. Bell, A new approach to the study of contact in the projective differential geometry of surfaces, *Duke Math. J.* **18** (1951), 689–695

[21] A. Berdon, *The Geometry of Discrete Groups*, Springer-Verlag, 1983

[22] M. Berger, *Geometry*, Springer-Verlag, 1994

[23] G. Birkhoff, *Dynamical Systems*, American Mathematical Society Colloq. Publ. 9, 1927

[24] I. Biswas, Differential operators on complex manifolds with a flat projective structure, *J. Math. Pures Appl.* **78** (1999), 1–26

[25] W. Blaschke, *Einfuhrung in die Geometrie der Waben*, Birkhäuser, 1955

[26] W. Blaschke, *Vorlesungen über Differentielgeometrie*, vol. 2, Springer-Verlag, 1923

[27] G. Bol, Über Nabelpunkte auf einer Einfläche, *Math. Z.* **49** (1943/44), 389–410

[28] G. Bol, Invarianten linearer differentialgleichungen, *Abh. Math. Sem. Univ. Hamburg* **16** (1949), 1–28

[29] G. Bol, *Projektive Differentialgeometrie*, Vandenhoeck & Ruprecht Göttingen, 1950

[30] H. Bos, C. Kers, F. Oort and D. Raven, Poncelet's closure theorem, *Expos. Math.* **5** (1987), 289–364

[31] R. Bott, On the characteristic classes of groups of diffeomorphisms, *Enseign. Math.* **23** (1977), 209–220

[32] S. Bouarroudj, Multi-dimensional Schwarzian derivative revisited, preprint arXiv:math.DG/0101056

[33] S. Bouarroudj and V. Ovsienko, Three cocycles on Diff(S^1) generalizing the Schwarzian derivative, *Int. Math. Res. Not.* (1998), No.1, 25–39

[34] S. Bouarroudj and V. Ovsienko, Schwarzian derivative related to modules of differential operators on a locally projective manifold, *Banach Center Publ.*, **51** (2000), 15–23

[35] K. Brown, *Cohomology of Groups*, Springer-Verlag, 1994

[36] H. Busemann, Problem IV: Desarguesian spaces, in *Mathematical Developments Arising from Hilbert Problems, Proceedings of the 28th Symposium on Pure Mathematics*, American Mathematical Society, 1976, 131–141

[37] E. Cartan, *Leçons sur la théorie des espaces à connexion projective*, Gauthier-Villars, 1937

[38] S. S. Chern, Web geometry, *Bull. Am. Math. Soc.* **6** (1982), 1–8

[39] S. Choi and W. Goldman, The classification of real projective structures on compact surfaces, *Bull. Am. Math. Soc.* **34** (1997), 161–171

[40] S. Choi and W. Goldman, Deformation space of convex $\mathbb{RP}^2$ structures on 2-orbifolds, preprint arXiv:math/0107193

[41] S. Choi and H. Lee, Geometric structures on manifolds and holonomy-invariant metrics, *Forum Math.* **9** (1997), 247–256

[42] K-S. Chou and X-P. Zhu, *The Curve Shortening Problem*, Chapman & Hall/CRC, 2001

[43] H. Cohen, Sums involving the values at negative integers of L-functions of quadratic characters, *Math. Ann.* **217** (1975), 271–285

[44] P. Cohen, Y. Manin and D. Zagier, Automorphic pseudodifferential operators, in A. S. Fokas and I. M. Gelfand (eds), *Algebraic Aspects of Integrable Systems*, Progress in Nonlinear Differential Equations and their Applications 26, Birkhäuser, 1997, 17–47

[45] R. Connely, Rigidity, in P. M. Gruber and J. M. Wills (eds), *Handbook of Convex Geometry*, North-Holland, 1993, 223–272

[46] W. Coppel, *Disconjugacy*, Lecture Notes in Mathematics 220, Springer-Verlag, 1971

[47] G. Darboux, *Léçons sur la théorie des surfaces*, Gauthier-Villars, 1896

[48] W. Degen, Projektive Differentialgeometrie, in O. Giering, (ed.) *et al.*, *Geometrie und ihre Anwendungen*, Carl Hanser Verlag, 1994, 319–374

[49] L. Dickey, *Soliton Equations and Hamiltonian Systems*, World Scientific, 1991

[50] J. Douglas, Solution to the inverse problem of the calculus of variations, *Trans. Am. Math. Soc.* **50** (1941), 71–128

[51] V. Drinfeld and V. Sokolov, Lie algebras and equations of Korteweg-de Vries type, in *Current Problems in Mathematics*, vol. 24, Itogi Nauki i Tekhniki, Akad. Nauk SSSR, 1984, 81–180

[52] B. Dubrovin, A. Fomenko and S. Novikov, *Modern Geometry – Methods and Applications. Part I. The geometry of surfaces, transformation groups, and fields*, Springer-Verlag, 1992

[53] C. Duval and L. Guieu, The Virasoro group and Lorentzian surfaces: the hyperboloid of one sheet, *J. Geom. Phys.* **33** (2000), 103–127

[54] C. Duval, P. Lecomte and V. Ovsienko, Conformally equivariant quantization: existence and uniqueness, *Ann. Inst. Fourier.* **49** (1999), 1999–2029

[55] C. Duval and V. Ovsienko, Space of second order linear differential operators as a module over the Lie algebra of vector fields, *Adv. Math.* **132** (1997), 316–333

[56] C. Duval and V. Ovsienko, Schwarzian derivative and Lorentzian wordlines, *Funct. Anal. Appl.* **34** (2000), 135–137

[57] C. Duval and V. Ovsienko, Projectively equivariant quantization and symbol calculus: noncommutative hypergeometric functions, *Lett. Math. Phys.* **57** (2001), 61–67

[58] L. Eisenhart, Spaces with corresponding paths, *Proc. Nat. Acad. Sci. USA* **8** (1922), 233–238

[59] C. Ehresmann, Sur les espaces localement homogènes, *Enseign. Math.* **35** (1936), 317–333

[60] A. Eremenko and D. Novikov, Oscillation of Fourier integrals with a spectral gap, *J. Math. Pure Appl.* **83** (2004), 313–365

[61] F. Fabricius-Bjerre, On the double tangents of plane curves, *Math. Scand.* **11** (1962), 113–116

[62] G. Faltings, Real projective structures on Riemann surfaces, *Compositio Math.* **48** (1983), 223–269

[63] E. Ferapontov, Stationary Veselov-Novikov equation and isothermally asymptotic surfaces in projective differential geometry, *Diff. Geom. Appl.* **11** (1999), 117–128

[64] E. Ferapontov, Integrable systems in projective differential geometry, *Kyushu J. Math.* **54** (2000), 183–215
[65] B. Feigin and D. Fuchs, Homology of the Lie algebra of vector fields on the line, *Func. Anal. Appl.* **14** (1980), 201–212
[66] E. Ferrand, On the Bennequin invariant and the geometry of wave fronts, *Geom. Dedicata* **65** (1997), 219–245
[67] M. Flato, A. Lichnerowicz, D. Sternheimer, Déformations 1-différentiables des algèbres de Lie attachées à une variété symplectique ou de contact, *Compositio Math.* **31** (1975), 47–82
[68] V. Fock, Dual Teichmüller spaces, preprint arXiv:math.dg-ga/9702018
[69] V. Fock and A. Goncharov, Moduli spaces of local systems and higher Teichmüller theory, preprint arXiv:math.AG/0311149
[70] V. Fock and A. Rosly, *Poisson Structure on Moduli of Flat Connections on Riemann Surfaces and r-matrix*, American Mathematical Society Translations, ser. 2, vol. 191, 1999, 67–86
[71] G. Fubini and E. Čech, *Introduction à la géométrie projective différentielle des surfaces*, Gauthier-Villars, 1931
[72] D. Fuchs, *Cohomology of Infinite-dimensional Lie Algebras*, Consultants Bureau, 1987
[73] M. Ganesh, Projective structures on compact complex manifolds – geometric theory, *J. Indian Math. Soc.* **42** (1978), 67–84
[74] H. Gargoubi, Sur la géométrie de l'espace des opérateurs différentiels linéaires sur $\mathbb{R}$, *Bull. Soc. R. Sci. Liège* **69** (2000), 21–47
[75] H. Gargoubi and V. Ovsienko, Space of linear differential operators on the real line as a module over the Lie algebra of vector fields, *Int. Math. Res. Not.* (1996) No. 5, 235–251
[76] H. Gargoubi and V. Ovsienko, Modules of differential operators on the line, *Func. Anal. Appl.* **35** (2001), 13–18
[77] I. Gelfand and L. Dikii, Fractional powers of operators, and Hamiltonian systems, *Func. Anal. Appl.* **10**:4 (1976), 13–29
[78] I. Gelfand and L. Dikii, A family of Hamiltonian structures connected with integrable nonlinear differential equations, Akad. Nauk SSSR Inst. Prikl. Mat. Preprint No. 136 (1978)
[79] E. Ghys, Cercles osculateurs et géométrie lorentzienne. Talk at the journée inaugurale du CMI, Marseille, February 1995
[80] E. Ghys, L'invariant de Godbillon-Vey, seminar, Bourbaki, *Astérisque* **177–178** (1989), 155–181
[81] C. Godbillon, *Feuilletages: études géometriques*, Birkhäuser, 1991
[82] J. Goldberg, *Theory of Multidimensional (n+1)-webs*, Kluwer, 1988
[83] W. Goldman, Projective geometry on manifolds, unpublished lecture notes, available from W. Goldman's website
[84] W. Goldman, *Geometric Structures on Manifolds and Varieties of Representations*, Contemporary Mathematics 74, American Mathematical Society, 1988, 169–198
[85] R. Goodman, N. Wallach, *Representations and Invariants of the Classical Groups. Encyclopedia of Mathematics and its Applications*, vol. **68** Cambridge University Press, 1998
[86] P. Gordan, *Invariantentheorie*, Teubner, 1887

[87] P. Grozman, Classification of bilinear invariant operators on tensor fields, *Func. Anal. Appl.* **14**:2 (1980), 58–59

[88] L. Guieu, E. Mourre and V. Ovsienko, *Theorem on Six Vertices of a Plane Curve via the Sturm Theory*, Arnold-Gelfand Mathematical Seminars, Birkhäuser, 1997, 257–266

[89] L. Guieu and V. Ovsienko, Structures symplectiques sur les espaces de courbes projectives et affines, *J. Geom. Phys.* **16** (1995), 120–148

[90] L. Guieu and C. Roger, *Groupe et algèbre de Virasoro*, **to appear**

[91] V. Guillemin, B. Kostant and S. Sternberg, Douglas' solution of the Plateau problem, *Proc. Nat. Acad. Sci. USA* **85** (1988), 3277–3278

[92] M. Hachtroudi, *Les espaces d'éléments a connexion projective normale*, Hermann, 1937

[93] H. Hamburger, Beweis einer Carathéodoryschen Vermutung I, *Ann. Math.* **41** (1940), 63–86

[94] H. Hamburger, Beweis einer Carathéodoryschen Vermutung II, *Acta Math.* **73** (1941), 175–228

[95] H. Hamburger, Beweis einer Carathéodoryschen Vermutung III, *Acta Math.* **73** (1941), 229–332

[96] S. Hansoul and P. Lecomte, Affine representations of Lie algebras and geometric interpretation in the case of smoooth manifolds, preprint arXiv:math.DG/0307261

[97] N. Hitchin, Lie groups and Teichmuller space, *Topology* **31** (1992), 449–473

[98] H. Hofer and E. Zehnder, *Symplectic Invariants and Hamiltonian Dynamics*, Birkhäuser, 1994

[99] A. Hurwitz, Uber die Fourierschen Konstanten integrierbarer Funktionen, *Math. Ann.* **57** (1903), 425–446

[100] V. Ivanov, The analytic Carathéodory conjecture, *Siberian Math. J.* **43** (2002), 251–322

[101] T. Ivey and J. Landsberg, *Cartan for Beginners: Differential Geometry via Moving Frames and Exterior Differential Systems*, American Mathematical Society, 2003

[102] H. Jacobowitz, Surfaces in RP^3, *Duke Math. J.* **46** (1979), 183–201

[103] S. Janson and J. Peetre, A new generalization of Hankel operators (the case of higher weights), *Math. Nachr.* **132** (1987), 313–328

[104] S. Karlin and W. Studden, *Tchebycheff Systems: With Applications in Analysis and Statistics*, John Wiley, 1966

[105] G. Katriel, From Rolle's theorem to the Sturm-Hurwitz theorem, preprint arXiv:math.CA/0308159

[106] B. Khesin and V. Ovsienko, Symplectic leaves of the Gelfand-Dikiï brackets and homotopy classes of nonflattening curves, *Func. Anal. Appl.* **24**:1 (1990), 33–40

[107] B. Khesin and B. Shapiro, Nondegenerate curves on S^2 and orbit classification of the Zamolodchikov algebra, *Comm. Math. Phys.* **145** (1992), 357–362

[108] A. Khovanskii and D. Novikov, Convex-concave body in $\mathbb{RP}^3$ contains a line, *Geom. Funct. Anal.* **13** (2003), 1082–1119

[109] A. Khovanskii and D. Novikov, On affine hypersurfaces with everywhere nondegenerate second quadratic form, preprint arXiv:math.DG/0203202

[110] A. Khovanskii and D. Novikov, L-convex-concave sets in real projective space and L-duality, *Moscow Math. J.* **3** (2003), 1013–1038

[111] R. Kirby and M. Scharlemann, Eight faces of the Poincaré homology 3-sphere, in *Geometric Topology*, Academic Press, 1979, 113–146

[112] A. Kirillov, *Elements of the Theory of Representations*, Springer-Verlag, 1976

[113] A. Kirillov, Local Lie algebras, *Russ. Math. Surv.* **31**:4 (1976), 57–76

[114] A. Kirillov, Invariant differential operators on geometric quantities, *Func. Anal. Appl.* **11**:2 (1977), 114–118

[115] A. Kirillov, The orbits of the group of diffeomorphisms of the circle, and local Lie superalgebras, *Func. Anal. Appl.* **15**:2 (1981), 75–76

[116] A. Kirillov, *Infinite Dimensional Lie Groups: Their Orbits, Invariants and Representations. The Geometry of Moments*, Lecture Notes in Mathematics 970, Springer-Verlag, 1982, 101–123

[117] F. Klein, *Lectures on the Icosahedron and the Solution of Equations of the Fifth Degree*, Teubner, 1913

[118] F. Klein, *Elementary Mathematics from an Advanced Standpoint*, Dover, 1948

[119] T. Klotz, On G. Bol's proof of Carathéodory's conjecture, *Comm. Pure Appl. Math.* **12** (1959), 277–311

[120] A. Kneser, *Bemerkungen über die Anzahl der Extreme der Krümmung auf geschlossenen Kurven und über vertwandte Fragen in einer nichteuklidischen Geometrie*, Festschrift H. Weber, 1912, 170–180

[121] S. Kobayashi, *Transformation Groups in Differential Geometry*, Springer-Verlag, 1972

[122] S. Kobayashi and T. Nagano, On projective connections, *J. Math. Mech.* **13** (1964), 215–235

[123] S. Kobayashi and K. Nomizu, *Foundations of Differential Geometry*, vols. 1, 2, John Wiley, 1963, 1969

[124] S. Kobayashi and T. Ochiai, Holomorphic projective structures on compact complex surfaces, *Math. Ann.* **249** (1980), 75–94

[125] J. Koszul, Homologie des complexes de formes différentielles d'ordre supérieur, *Ann. Sci. Ecole Normale Sup.* **7** (1974), 139–154

[126] N. Kuiper, Locally projective spaces of dimension one, *Michigan Math. J.* **2** (1953), 95–97

[127] F. Labourie, $\mathbb{RP}^2$-strictures et differentielles cubique holomorphe, preprint 1997

[128] J. Landsberg, *Algebraic Geometry and Projective Differential Geometry*, Seoul National University Lecture Notes Series 45, 1999

[129] E. Lane, *A Treatise on Projective Differential Geometry*, University of Chicago Press, 1942

[130] V. Lazutkin and T. Pankratova, Normal forms and versal deformations for Hill's equation, *Func. Anal. Appl.* **9**:4 (1975), 41–48

[131] P. Lecomte, Classification projective des espaces d'opérateurs différentiels agissant sur les densités, *C. R. Acad. Sci. Paris Sér. I Math.* **328** (1999), 287–290

[132] P. Lecomte, On the cohomology of $\mathrm{sl}(m+1,\mathbb{R})$ acting on differential operators and $\mathrm{sl}(m+1,\mathbb{R})$-equivariant symbol, *Indag. Math.*, **11** (2000), 95–114

[133] P. Lecomte and V. Ovsienko, Cohomology of the vector fields Lie algebra and modules of differential operators on a smooth manifold, *Compositio Math.* **124** (2000), 95–110

[134] P. Lecomte and V. Ovsienko, Projectively invariant symbol calculus, *Lett. Math. Phys.* **49** (1999), 173–196

[135] A. Li, U. Simon and G. Zhao, *Global Affine Differential Geometry*, de Gruyter, 1993

[136] A. Lichnerowicz, *Geometry and Cohomologies Associated with a Contact Manifold*, Lecture Notes in Mathematics 1045, Springer-Verlag, 1984, 104–116

[137] J. Little, Nondegenerate homotopies of curves on the unit 2-sphere, *J. Diff. Geom.* **4** (1970), 339–348

[138] J. Little, Third order nondegenerate homotopies of space curves, *J. Diff. Geom.* **5** (1971), 503–515

[139] J. Loftin, The compactification of the moduli space of convex $\mathbb{RP}^2$ surfaces, preprint arXiv:math.DG/0311052

[140] F. Magri, A simple model of the integrable Hamiltonian equations, *J. Math. Phys.* **19** (1978), 1156–1162

[141] G. Marí-Beffa and P. Olver, Differential invariants for parametrized projective surfaces, *Comm. Anal. Geom.* **7** (1999), 807–839

[142] Y. Martinezz-Maure, Les multihérissons et le théorème de Sturm-Hurwitz, *Arch. Math.* **80** (2003), 79–86

[143] P. Mathonet, Intertwining operators between some spaces of differential operators on a manifold, *Comm. Algebra* **27** (1999), 755–776

[144] V. Matveev and P. Topalov, Trajectory equivalence and corresponding integrals, *Reg. Chaotic Dyn.* **3** (1998), 30–45

[145] V. Matveev and P. Topalov, Dynamical and topological methods in theory of geodesically equivalent metrics, *Zap. Nauchn. Sem. POMI* **266** (2000), 155–168

[146] V. Matveev and P. Topalov, Geodesic equivalence of metrics as a particular case of the integrability of geodesic flows, *Theor. Math. Phys.* **123** (2000), 651–658

[147] V. Matveev and P. Topalov, Geodesic equivalence via integrability, *Geom. Dedicata* **96** (2003), 91–115

[148] D. McDuff and D. Salamon, *Introduction to Symplectic Topology*, Clarendon Press, 1995

[149] A. F. Möbius, Über die Grundformen der Linien der dritten Ordnung, Abhandlungen der Königl. Sächs. Gesellschaft der Wissenschaften, *Math.-Phys. Klasse* **1** (1852), 1–82

[150] J. Moser, Various aspects of integrable Hamiltonian systems, in *Dynamical Systems (CIME Summer School, Bressanone, 1978)*, Progress in Mathematics 8, Birkhäuser, 1980, 233–289

[151] J. Moser, Geometry of quadrics and spectral theory, in *The Chern Symposium 1979 (Proceedings of the International Symposium, Berkeley, California, 1979)*, 1979, Springer-Verlag, 1980, 147–188

[152] J. Moser, *Integrable Hamiltonian Systems and Spectral Theory*. Lezioni Fermiane, 1981

[153] J. Moser and A. Veselov, Discrete versions of some classical integrable systems and factorization of matrix polynomials, *Comm. Math. Phys.* **139** (1991), 217–243

[154] S. Mukhopadhyaya, New methods in the geometry of a plane arc, *Bull. Calcutta Math. Soc.* **1** (1909), 32–47

[155] R. Molzon and K. Mortensen, The Schwarzian derivative for maps between manifolds with complex projective connections, *Trans. Am. Math. Soc.* **348** (1996), 3015–3036

[156] K. Nomizu and T. Sasaki, *Affine Differential Geometry. Geometry of Affine Immersions*, Cambridge University Press, 1994

[157] P. Olver and J. Sanders, Transvectants, modular forms, and the Heisenberg algebra, *Adv. Appl. Math.* **25** (2000), 252–283

[158] H. Omori, Y. Maeda, N. Miyazaki and A. Yoshioka, Deformation quantizations of the Poisson algebra of Laurent polynomials, *Lett. Math. Phys.* **46** (1998), 171–180

[159] B. Osgood, Old and new on the Schwarzian derivative, in J. M. Heinonen (ed.), *Quasiconformal Mappings and Analysis*, Springer-Verlag, 1998, 275–308

[160] *Outside In*: A video about turning the sphere inside out. The Geometry Center, University of MN, A.K. Peters, 1994

[161] V. Ovsienko, Classification of third-order linear differential equations and of symplectic leaves of the Gelfand-Dikii bracket, *Math. Notes* **47** (1990), 465–470

[162] V. Ovsienko, Contact Virasoro algebra, *Func. Anal. Appl.* **24** (1990), 54–62

[163] V. Ovsienko, On the Denogardius great number and Hooke's law, in S. Tabachnikov (ed.), *Kvant Selecta: Algebra and Analysis*, vol. 2, American Mathematical Society, 1999, 153–159

[164] V. Ovsienko, Projective structures and contact forms, *Func. Anal. Appl.* **28** (1994), 187–197

[165] V. Ovsienko, Lagrange Schwarzian derivative and symplectic Sturm theory, *Ann. Fac. Sci. Toulouse* **2** (1993), 73–96

[166] V. Ovsienko, Selfadjoint differential operators and curves on a Lagrangian Grassmannian that are subordinate to a loop, *Math. Notes* **47** (1990), 270–275

[167] V. Ovsienko, Exotic deformation quantization, *J. Diff. Geom.* **45** (1997), 390–406

[168] O. Ovsienko and V. Ovsienko, Lie derivatives of order n on the line. Tensor meaning of the Gelfand-Dikiĭ bracket, in A. A. Kirillov (ed.) *Topics in Representation Theory*, Advances in Soviet Mathematics 2, American Mathematical Society, 1991, 221–231

[169] V. Ovsienko and C. Roger, Generalizations of Virasoro group and Virasoro algebra through extensions by modules of tensor-densities on S^1, *Indag. Math.* **9** (1998), 277–288

[170] V. Ovsienko and C. Roger, Deformations of Poisson brackets and extensions of Lie algebras of contact vector fields, *Russ. Math. Surv.* **47** (1992), 135–191

[171] V. Ovsienko and S. Tabachnikov, Sturm theory, Ghys theorem on zeroes of the Schwarzian derivative and flattening of Legendrian curves, *Selecta Math.* (N.S.) **2** (1996), 297–307

[172] V. Ovsienko and S. Tabachnikov, Projective geometry of polygons and discrete 4-vertex and 6-vertex theorems, *L'Enseign. Math.* **47** (2001), 3–19

[173] J. Peetre, Une caractérisation abstraite des opérateurs différentiels, *Math. Scand.* **7** (1959), 211–218 and **8** (1960), 116–120

[174] G. Pólya and G. Szegö, *Problems and Theorems in Analysis*, Springer-Verlag, 1978

[175] G. Pólya and N. Wiener, On the oscillation of the derivatives of a periodic function, *Trans. Am. Math. Soc.* **52** (1942), 249–256

[176] A. Pogorelov, *Hilbert's Fourth Problem*, John Wiley, 1979

[177] R. Rankin, The construction of automorphic forms from the derivatives of a given form, *J. Indian Math. Soc.* **20** (1956), 103–116

[178] C. Roger, Various aspects and generalizations of the Godbillon-Vey invariant, in B. P. Komrakov (ed.) *Lie Groups and Lie Algebras*, Mathematics and its Applications 433, Kluwer, 1998, 231–247

[179] A. Rudakov, Irreducible representations of infinite-dimensional Lie algebras of Cartan type, *Izv. Akad. Nauk SSSR Ser. Mat.* **38** (1974), 835–866

[180] H. Rund, *The Differential Geometry of Finsler Spaces*, Springer-Verlag, 1959

[181] L. Santalo, *Integral Geometry and Geometric Probability*, Addison-Wesley, 1976

[182] R. Schwartz, Pappus' theorem and the modular group, *Publ. Math. IHES* **78** (1993), 187–206

[183] R. Schwartz, Desargues theorem, dynamics, and hyperplane arrangements, *Geom. Dedicata* **87** (2001), 261–283

[184] V. Sedykh, Theorem of four support vertices of a polygonal line, *Func. Anal. Appl.* **30**:3 (1996), 216–218

[185] V. Sedykh, *Discrete Versions of Four-Vertex Theorem*, American Mathematical Society Translations, ser. 2, vol. 180, 1997, 197–207

[186] G. Segal, Unitary representations of some infinite-dimensional groups, *Comm. Math. Phys.* **80** (1981), 301–342

[187] B. Segre, *Some Properties of Differentiable Varieties and Transformations*, Springer-Verlag, 1957

[188] B. Segre, Alcune proprietà differenziali in grande delle curve chiuse sghembe, *Rend. Mat.* **1** (1968), 237–297

[189] M. Shapiro, The topology of the space of nondegenerate curves, *Russ. Acad. Sci. Izv. Math.* **43**:2 (1994), 291–310

[190] U. Simon, A. Schwenk-Schellschmidt and H. Viesel, Introduction to the affine differential geometry of hypersurfaces, Lecture notes, Science University, Tokyo, 1991

[191] U. Simon, Affine differential geometry, in *Handbook of Differential Geometry*, vol. 1, Elsevier, 2000, 905–962

[192] D. Singer, Diffeomorphisms of the circle and hyperbolic curvature, *Conform. Geom. Dyn.* **5** (2001), 1–5

[193] M. Spivak, *A Comprehensive Introduction to Differential Geometry*, Publish or Perish, 1979

[194] D. Spring. The golden age of immersion theory in topology: 1959–1973, preprint arXiv:math.GT/0307127

[195] C. Sturm, Sur une classe d'équations à différences partielles, *J. Math. Pures Appl.* **1** (1836), 373–444

[196] D. Sullivan and W. Thurston, Manifolds with canonical coordinate charts: some examples, *L'Enseign. Math.* **29** (1983), 15–25

[197] Z. Szabo, Hilbert's fourth problem I, *Adv. Math.* **59** (1986), 185–301

[198] S. Tabachnikov, *Billiards,* Société Mathématique de France, 1995

[199] S. Tabachnikov, Calculation of the Bennequin invariant of a Legendrian curve by the geometry of its front, *Func. Anal. Appl.* **22**:3 (1988), 89–90

[200] S. Tabachnikov, Around four vertices, *Russ. Math. Surv.* **45**:1 (1990), 191–192

[201] S. Tabachnikov, Geometry of Lagrangian and Legendrian 2-web, Diff. Geom. Appl. **3** (1993), 265–284

[202] S. Tabachnikov, Projective connections, group Vey cocycle and deformation quantization, *Int. Math. Res. Not.* (1996) No. 14, 705–722

[203] S. Tabachnikov, Exact transverse line fields and projective billiards in a ball, *Geom. Func. Anal.* **7** (1997), 594–608

[204] S. Tabachnikov, *On Zeroes of the Schwarzian Derivative*, American Mathematical Society Translations, ser. 2, vol. 180, 1997, 229–239

[205] S. Tabachnikov, Introducing projective billiards, *Ergodic Theory Dynam. Syst.* **17** (1997), 957–976

[206] S. Tabachnikov, Parameterized curves, Minkowski caustics, Minkowski vertices and conservative line fields, *L'Enseign. Math.* **43** (1997), 3–26

[207] S. Tabachnikov, Geometry of exact transverse line fields and projective billiards, in S. Tabachnikov (ed.) *Differential and Symplectic Topology of Knots and Curves*, American Mathematical Society Translations, ser. 2, vol. 190, 1999, 131–152

[208] S. Tabachnikov, Remarks on the geometry of exact transverse line fields, in A. Astashkevich and S. Tabachnikov (ed.), *Differential Topology, Infinite-dimensional Lie Algebras and Applications*, American Mathematical Society Translations, ser. 2, vol. 194, 1999, 247–260

[209] S. Tabachnikov, Projectively equivalent metrics, exact transverse line fields and the geodesic flow on the ellipsoid, *Comment. Math. Helv.* **74** (1999), 306–321

[210] S. Tabachnikov, A four vertex theorem for polygons, *Am. Math. Monthly* **107** (2000), 830–833

[211] S. Tabachnikov, Dual billiards in the hyperbolic plane, *Nonlinearity* **15** (2002), 1051–1072

[212] S. Tabachnikov, Ellipsoids, complete integrability and hyperbolic geometry, *Moscow Math. J.* **2** (2002), 185–198

[213] S. Tabachnikov, Remarks on magnetic flows and magnetic billiards, Finsler metrics and a magnetic analog of Hilbert's fourth problem, preprint arXiv:math.DG/0302288

[214] I. Tamura, *The Topology of Foliations*, Iwanami Shoten, 1976

[215] T. Thomas, On the projective and equiprojective geometry of paths, *Proc. Nat. Acad. Sci. USA* **11** (1925), 199–203

[216] G. Thorbergsson and M. Umehara, A unified approach to the four vertex theorem II, in S. Tabachnikov (ed.) *Differential and Symplectic Topology of Knots and Curves*, American Mathematical Society Translations, ser. 2, vol. 190, 1999, 229–252

[217] G. Thorbergsson and M. Umehara, Sextactic points on a simple closed curve, *Nagoya Math. J.* **167** (2002), 55–94

[218] G. Thorbergsson and M. Umehara, A global theory of flexes of periodic functions, *Nagoya Math. J.* **173** (2004), 85–138

[219] W. Thurston, Noncobordant foliations of S^3, *Bull. Am. Math. Soc.* **78** (1972), 511–514

[220] W. Thurston, *Three-Dimensional Geometry and Topology*, vol. 1, Princeton University Press, 1997

[221] Ch. Titus, A proof of a conjcture of Loewner and the conjecture of Carathéodory on umbilic points, *Acta Math.* **131** (1973), 43–77

[222] A. Tjurin, Periods of quadratic differentials, *Russ. Math. Surv.* **33** (1978), 169–221

[223] T. Tsujishita, *Continuous Cohomology of the Lie Algebra of Vector Fields*, *Memoirs Am. Math. Soc.*, 253, 1981

[224] M. Umehara, A unified approach to the four vertex theorem I, in S. Tabachnikov (ed.), *Differential and Symplectic Topology of Knots and Curves*, American Mathematical Society Translations, ser. 2, vol. 190, 1999, 185–228

[225] O. Veblen, Projective and affine geometry of paths, *Proc. Nat. Acad. Sci. USA* **8** (1922), 347–352

[226] A. Veselov, Integrable maps, *Russ. Math. Surv.* **46**:5 (1991), 3–45

[227] A. Veselov, Confocal surfaces and integrable billiards on the sphere and in the Lobachevsky space, *J. Geom. Phys.* **7** (1990), 81–107

[228] B. Wegner, On the evolutes of piecewise linear curves in the plane, *Rad. Hrvat. Acad. Znan. Um.* **467** (1994), 1–16

[229] J. Weiner, Global properties of spherical curves, *J. Diff. Geom.* **12** (1977), 425–438

[230] A. Weinstein, The local structure of Poisson manifolds, *J. Diff. Geom.* **18** (1983), 523–557

[231] E. Wilczynski, *Projective Differential Geometry of Curves and Ruled Surfaces*, Teubner, 1906

[232] E. Wilczynski, Projective differential geometry of curved surfaces I–V, *Trans. Am. Math. Soc.* **8** (1907), 233–260; **9** (1908), 79–120, 293–315; **10** (1909), 176–200, 279–296

[233] J. M. C. Whitehead, Locally homogeneous spaces in differential geometry, *Ann. Math.* **33** (1932), 681–687

[234] M. Yoshida, Canonical forms of some systems of linear partial differential equations, *Proc. Japan Acad.* **52** (1976), 473–476

Index

www.ingramcontent.com/pod-product-compliance
Lightning Source LLC
LaVergne TN
LVHW101500200926
848483LV00001B/181
9780521831864